£3 -

PERGAMON INTERNATIONAL LIBRARY
of Science, Technology, Engineering & Social Studies

The 1000-volume original paperback library in aid of education,
industrial training and the enjoyment of leisure

Publisher: Robert Maxwell, M.C.

The Computation of Style

AN INTRODUCTION TO STATISTICS FOR STUDENTS OF LITERATURE AND HUMANITIES

Other titles of interest

HOWARD-HILL, T. H.
Literary Concordances

MYERS, R.
A Dictionary of Literature in the English Language from Chaucer to 1970

TAYLOR, T. J.
Linguistic Theory and Structural Stylistics

A related journal

LANGUAGE & COMMUNICATION*
An interdisciplinary journal dedicated to the investigation of language
and its communicational function

Editor: Roy HARRIS, *University of Oxford*

Free specimen copy available on request.

The Computation of Style

AN INTRODUCTION TO STATISTICS FOR STUDENTS OF LITERATURE AND HUMANITIES

by

ANTHONY KENNY

Master of Balliol College, Oxford

PERGAMON PRESS

OXFORD · NEW YORK · TORONTO · SYDNEY · PARIS · FRANKFURT

U.K.	Pergamon Press Ltd., Headington Hill Hall, Oxford OX3 0BW, England
U.S.A.	Pergamon Press Inc., Maxwell House, Fairview Park, Elmsford, New York 10523, U.S.A.
CANADA	Pergamon Press Canada Ltd., Suite 104, 150 Consumers Rd., Willowdale, Ontario M2J 1P9, Canada
AUSTRALIA	Pergamon Press (Aust.) Pty. Ltd., P.O. Box 544, Potts Point, N.S.W. 2011, Australia
FRANCE	Pergamon Press SARL, 24 rue des Ecoles, 75240 Paris, Cedex 05, France
FEDERAL REPUBLIC OF GERMANY	Pergamon Press GmbH, 6242 Kronberg-Taunus, Hammerweg 6, Federal Republic of Germany

First edition 1982

Library of Congress Cataloguing in Publication Data

Kenny, Anthony John Patrick.
The computation of style.
(Pergamon international library of science, technology, engineering, and social sciences)
1. Literary research—Statistical methods. 2. Statistics. 3. Style, Literary. I. Title. II. Series. PN73.K46
1982 807.2 81-19221 AACR2

British Library Cataloguing in Publication Data

Kenny, Anthony
The computation of style.
1. English literature—Style 2. Electronic digital computers
I. Title
820.9 PR57

ISBN 0-08-024282-0 (Hardcover)
ISBN 0-08-024281-2 (Flexicover)

Printed and bound in Great Britain by A. Wheaton & Co., Exeter

Preface

THIS book is intended as an elementary introduction to statistics for those who wish to make use of statistical techniques in the study of literature. There are many elementary textbooks of statistics on the market, but they are all aimed at students in the natural or social sciences. In general they are unsuitable for scholars in the humanities because they presuppose too great a facility in mathematics and because they fail to emphasize or illustrate the application of statistical methods to literary material. To the best of my knowledge there exists no elementary statistical text in English written with the needs and interests of the literary student specifically in mind.

The reader of the present book is not assumed to have any mathematical competence beyond a rusty memory of junior school arithmetic and algebra. The illustrations in the text have been carefully chosen to make the calculations involved as simple as possible. The exercises have been carefully selected so that the calculation of the relevant statistics will be possible for anyone who can remember how to do long division. (They will not involve, for instance, the extraction of difficult square roots.) All these problems can be solved without the aid of a slide-rule, pocket calculator, or computer. To facilitate the solutions in this way the literary texts have occasionally been slightly doctored, and the statistics obtained are not usually of any literary interest except as illustrations of technique. Had the examples been less artificially chosen, the calculation would have been cumbrous without an aid such as a calculator, but the results would have been of greater interest from a literary point of view.

Any reader who proposes to become a serious student of literary statistics will have to acquire, sooner or later, a suitable calculator. However, the book is so designed that no one should need a calculator to follow the argument of the text or work out the exercises.

v

Calculators are now available at reasonable prices which not only facilitate the computation of statistics but also dispense with the consultation of statistical tables. For this reason the number of tables included in this volume has been kept down to the minimum necessary to introduce the various techniques involved to the reader who has not yet acquired a calculator.

The present book covers approximately the same ground as that in any first-year statistics textbook: it differs from other books in that it places emphasis on those techniques which are most useful in literary contexts and in the choice of examples drawn almost entirely from literary and linguistic material.

Like the readers for whom this book is intended, I came to statistics as a mathematical ignoramus with a purely humanistic background. I therefore owe a great deal to the text-books of other writers from which I have learnt. In acknowledging my debt to them, I shall also explain that they still seemed to leave room for a textbook aimed at a specifically humanistic audience.

The three textbooks I found most helpful were Donald Ary and Lucy Cheser Jacobs, *Introduction to Statistics: Purposes and Procedures*; Evelyn Caulcott, *Significance Tests*; George W. Snedecor and William G. Cochran, *Statistical Methods*. The first is the most clearly written and easily intelligible introduction to descriptive statistics I have encountered; the second gives the fullest coverage, at an elementary level, to the branch of inferential statistics which is most useful to the literary scholar. Both books, however, are aimed at the social scientist and draw their examples from sociology and economics and related fields. The third is considerably more advanced than the first two and progresses much further than the present volume. It contains much information that is useful to the literary statistician, but it is difficult going for someone without a mathematical background, and it is particularly oriented towards the agriculture, medical and biological sciences.

Charles Muller's *Initiation aux Méthodes de la Statistique Linguistique* is aimed at very much the same audience as the present book. However, there were a number of reasons why it seemed preferable to write a new book rather than to translate Muller for an English audience. In the first place, Muller's examples, naturally enough, are drawn mainly from French literature and in particular from his own work on Corneille. In the second place, the book lacks exercises and would be difficult to use without the separately published *Exercises d'Application* as an elementary textbook. But any reader of the present book will be able to read Muller with pleasure and profit.

The present book, while introducing the reader to the use of statistical techniques in literary context, does not attempt any general evaluation of the merits of statistical stylometry; rather it attempts to put the reader in a position to make such evaluations for himself. There are a number of books canvassing the value of stylometry: a recent and lively one is A. Q. Morton's *Literary Detection*. Morton is an energetic and imaginative pioneer in the field, and most of his books contain a brief introduction to statistical methods; but in my view his presentation of these methods is too brisk and allusive to provide an initiation for the innumerate beginner.

I am very much indebted to N. D. Thomson who eliminated many errors from an early draft of this book. Responsibility for errors which remain is, of course, my own.

ANTHONY KENNY
Balliol, January 1981

Contents

1

The Statistical Study of Literary Style

THE beginning of the statistical study of style in modern times is commonly dated to 1851 when Augustus de Morgan suggested that disputes about the authenticity of some of the writings of St Paul might be settled by the measurement of the length, measured in letters, of the words used in the various Epistles. The first person actually to test the hypothesis that word-length might be a distinguishing characteristic of writers was an American geophysicist called T. C. Mendenhall, who expounded his ideas in a popular journal, *Science*, in 1887. As a spectroscope analyses a beam of light into a spectrum characteristic of an element, he wrote, so 'it is proposed to analyse a composition by forming what may be called a "word spectrum" which shall be a graphic representation of words according to their length and to their relative frequency of occurrence'. He hoped that each author might be found to be as constant in his habits as each element in its properties. 'If it can be proved that the word spectrum... of *David Copperfield* is identical with that of *Oliver Twist*, of *Barnaby Rudge*, of *Great Expectations*... and that it differs sensibly from that of *Vanity Fair*, or *Eugene Aram*, or *Robinson Crusoe*... then the conclusion will be tolerably certain that when it appears it means Dickens.'

Mendenhall took several authors and constructed their word-spectra, or what would nowadays be called frequency distributions of word-length. He studied a number of samples of *Oliver Twist* and found great similarities between them. Three-letter words were always the commonest; this was true also in 10,000 words studied from *Vanity Fair*. But when two samples of J. S. Mill's work were investigated, it was found that the two-letter words were the most popular.

The average length of word in Mill, however, was longer than that in either Dickens or Thackeray: it was 4.775, compared with Dickens' 4.312 and Thackeray's 4.481. It was longer also than either part of the Authorized Version, where the average was 4.604 for the Old Testament and 4.625 for the New. De Morgan seems to have thought that the average word-length by itself might be an indication of authorship; but Mendenhall's studies showed that texts with the same average word-length might possess different spectra. The comparison between Mill and other authors provided a striking example of this.

It was by comparing the whole spectrum of word-length preferences that Mendenhall hoped to offer a scientific solution to disputes about authorship. His most ambitious attempt in this direction was a study of the Shakespeare–Bacon controversy in the *Popular Science Monthly* in 1901. With a grant from a Boston philanthropist, Mendenhall employed two secretaries and a counting machine to analyse some 400,000 words of Shakespeare, 200,000 words of Bacon and quantities of text from authors of other periods. He found that Shakespeare's 'characteristic curve' was very consistent from sample to sample, whether poetry or prose; it had unusual features. 'Shakespeare's vocabulary consisted of words whose average length was a trifle below four letters, less than that of any writer of English before studied; and his word of greatest frequency was the *four-letter word*, a thing never met with before.' These characteristics marked Shakespeare out from most of his contemporaries just as sharply as from the 19th-century authors previously studied. In particular his characteristic curve differed from that of Bacon. But, to his consternation, Mendenhall discovered that 'in the characteristic curve of his plays Christopher Marlowe agrees with Shakespeare about as well as Shakespeare agrees with himself'.

While Mendenhall had been using stylistic statistics in the attempt to solve attribution problems in English, European scholars had been developing stylometric techniques for Greek in order to settle the disputed chronology of Plato's dialogues. Lewis Campbell, a Balliol classicist who became Professor of Greek at St Andrews, offered in his 1867 edition of the *Sophist* and *Politicus* a battery of stylistic tests which he believed to be indicators of relative date: word order, rhythm, the avoidance of hiatus, and especially 'originality of vocabulary' as measured by the frequency of *hapax legomena* or once-occurring words, which he noted and counted carefully. He concluded that the *Sophist* and *Politicus* belonged to a late period in Plato's life, after the *Republic* and close to the *Laws*. His work passed unnoticed for nearly 30 years, but the German philologist Ritter reached similar conclusions by similar methods in 1888.

The work of these and other Platonic scholars was magisterially synthesized by the Polish scholar W. Lutosławski in his work of 1897, *The Origin and Growth of Plato's Logic*. Lutosławski lists 500 different facets of style which he believed could be used as stylometric indicators, and he was insistent on the importance of complete enumeration and numerical precision. For instance among the features he listed are the following:

318. Answers denoting subjective assent less than once in 60 answers.
325. Superlatives in affirmative answers more than half as frequent as positives, but not prevailing over positives.
378. Interrogations by means of *ara* between 15 and 24% of all interrogations.
412. *Peri* placed after the word to which it belongs forming more than 20% of all occurrences of *peri*.

He went so far as to formulate an explicit hypothesis which governed his stylometric investigations. He called it the 'Law of stylistic affinity', and he enunciated it thus:

> Of two works of the same author and of the same size, that is nearer in time to a third, which shares with it the greater number of stylistic peculiarities, provided that their different importance is taken into account, and that the number of observed peculiarities is sufficient to determine the stylistic character of all the three works.

The total number of peculiarities to be considered, he reckoned, should be not less than 150 in a dialogue of ordinary length.

The stylistic features studied by these Platonic scholars were much more subtle and sophisticated than the crude measure of word-length investigated by Mendenhall. On the other hand, Mendenhall had a surer grasp of statistical principles than Lutosławski, despite the impressiveness of some of the latter's numerical tables. As for their principle theoretical assumptions, Mendenhall's assumption that everyone has a uniquely characteristic word-spectrum, and Lutosławski's confident enunciation of the 'Law of stylistic affinity' look equally rash from the viewpoint of nearly a century of hindsight.

Campbell and Lutosławski were concerned with quantifiable variations of style as an indicator of chronology in a single author. An American philologist, L. A. Sherman of Nebraska, in papers written in 1888 and 1892, raised the possibility of studying by similar methods the evolution of a language as a whole. He claimed that literary English over the centuries had been developing from a heavy classical diction to a clearer and more natural form of expression. To prove this he and his pupils studied the average sentence-length in 300-sentence samples from a variety of authors and measured the proportion of simple to compound sentences in their work.

In the present century sentence-length has been studied rather as a distinctive characteristic of individual writers. The Cambridge statistician Udny Yule gave counts for sentence-lengths from Bacon, Coleridge and Macaulay, and was able to show striking differences between these authors. He studied a number of authorship attribution problems, especially the question whether the Latin medieval classic, *The Imitation of Christ*, was written by Thomas à Kempis or Jean Gerson. He found difficulty in the use of sentence-length as a discriminator in his preliminary study in 1938, and when he returned to the topic in 1944 it was through the study of vocabulary that he attempted to settle the authorship. He devised a complicated measure of vocabulary richness, the 'Characteristic K', which a number of later scholars have adapted and modified. He made statistical comparisons between the *Imitation* and the undisputed works of Gerson and à Kempis, and concluded that à Kempis was more likely to have been the author. The work in which he published his results, *The Statistical Study of Literary Vocabulary*, was the first full-length volume on the topic by a statistician and remains a classic in the field.

Between Mendenhall and Yule the discipline of statistics itself had developed enormously. When Mendenhall wrote there were no well-established techniques for deciding when differences between different phenomena were significant differences and when they were merely the result of random variation such as occurs between one deal and the next in a card game. It was not yet clearly understood when generalizations about large aggregates could safely be based on the study of comparatively small samples. The sociologist Rowntree, studying poverty in York in 1899, surveyed every single working-class household, just as Mendenhall had felt obliged to count nearly two million words for his Shakespearean study. Mendenhall's studies have been revived, and improved upon, in the present century. The statistician responsible (C. B. Williams, *Style and Vocabulary*, 1970) says that 'it is almost certain that conclusions of quite sufficient accuracy could have been obtained with one-tenth of the labour by a good system of sampling'. When Rowntree's final survey of York was carried out in 1950 the theory of sampling had developed so far that it was not necessary to survey more than 11% of the working-class households.

Since the Second World War the labour of the statistical investigation of style has been enormously reduced in another way—through the development of electronic computers. First, and obviously, the computer can perform the necessary statistical calculations more speedily and accurately than any human investigator. But more importantly the computer can assist in the provision of the data on which any statistical calculation must be based. Once a literary text has been fed into a computer, it is comparatively simple to measure word- and sentence-lengths of large portions of text and to provide the frequency counts and concordances on which vocabulary studies must be based. There now exist computer languages specially designed to facilitate the writing of programs to perform such operations; and there also exist packages of predesigned routines for concordance-making to save the student the labour even of writing his own programs. Inputting a text in natural language to a computer is far from being a trivial or inexpensive task; but in recent decades libraries of machine-readable texts have been built up in a number of centres.

It is not surprising, then, that since the war there has been a great increase in the number of quantitative studies of literary texts. The value of these studies has been very uneven, but they include a number which have been widely influential, and a few which can be regarded as models of procedure. Most of the stylistic features studied are ones which had already caught the attention of 19th-century scholars; but the development of statistical techniques, and the increasing availability of data for comparison, are gradually building the study into a discipline in its own right and with its own standards.

I will give a brief account of a number of statistical studies of style undertaken since the war, not with any view of producing a complete survey, but to illustrate the types of problems that have been studied and the kinds of methods used.

The question of the authenticity of the Epistles in the Pauline corpus has occupied New Testament scholars for a long time. W. C. Wake (1948 and 1957) and after him A. Q. Morton, studied the length of sentences in the Epistles and made extensive comparisons with sentence-lengths in other Greek authors. They concluded that the variability in sentence-lengths between Epistles was beyond parallel in any of the other authors studied. On the basis of this, and of the occurrences of a number of frequent conjunctions and particles, Morton, in his *Paul, the Man and the Myth* (1966), argued that Romans, I and II Corinthians, and Galatians formed a homogeneous group which could be attributed to the Apostle Paul; between this group and the other Epistles there were a large number of significant differences, some of them larger than any differences known to exist in the writings of any Greek author, regardless of literary form or variation over time. The other Epistles, Morton suggested, might come from as many as six different hands. Morton has returned to the topic in a number of subsequent studies, offering fresh evidence from different stylistic features, all pointing in the direction of the same conclusion. A striking feature of Morton's work is that it reaches, by totally different methods, conclusions first suggested by F. C. Baur at Tübingen in the early 19th century. A weakness in his argument is that the methods he used reveal substantial anomalies in II Corinthians and in Romans itself, anomalies which he has not been altogether successful in explaining away. But New Testament scholars who wish to maintain the unity of the Pauline corpus have not yet made an adequate response to the impressive weight of evidence he has amassed.

In the early 1960s there appeared two studies of English texts which are widely regarded as models of the statistical investigation of style: Alvar Ellegård's study of the *Junius Letters*, and the work of Frederick Mosteller and David Wallace on the *Federalist Papers*.

The *Junius Letters* are a series of political pamphlets written in 1769–1772 whose authorship is one of the mysteries of English literary history. Some 40 people have been suggested as possible authors; since the middle of the 19th century Sir Philip Francis has been generally regarded as the most likely candidate. Ellegård, a Swedish literary historian, took up the problem in 1962 in a pair of books which combined a careful survey of the traditional historical and biographical evidence, with a novel 'statistico-linguistic' investigation. In his book *A Statistical Method for Determining Authorship* he presented a list of nearly 500 words and expressions which he had found to be characteristic of Junius—either positively, being used more frequently by Junius than his contemporaries, or negatively, being less favoured by him. He used this list, plus an examination of the choices made by Junius between some 50 pairs or triplets of approximate synonyms (e.g. *on* vs *upon*; *kind* vs *sort*), to calculate the characteristics which Junius shared with each of the candidates for authorship. It turned out that there was a remarkable accord between the habits of Junius and those of Sir Philip Francis. If we assume that no more than 300 people sufficiently fit the relevant biographical data to be candidates for authorship, then we can be 99% certain, Ellegård concluded, that Sir Philip Francis was Junius. Ellegård's book is particularly valuable for the reflections it contains on the general methodology of statistical tests in literary contexts.

To determine the habits which marked Junius as different from his contemporaries, Ellegård had to study more than a million words of contemporary works. The problem attacked by Mosteller and Wallace was a more tractable one, since there were only two candidates for the authorship of the texts in question. The *Federalist Papers* are a series of articles published in 1787 and 1788 to persuade the citizens of New York to ratify the Constitution. The papers were written by Jay, Hamilton, and Madison; the authorship of most was known, but 12 out of the 88 were contested between Hamilton and Madison. The papers of known authorship here provided the material from which features discriminating the two candidates could be extracted.

Mosteller and Wallace, like Ellegård, looked for 'marker words', words which were particular likes or dislikes of the competing authors. Madison and Hamilton, for instance, differed consistently in their choice between 'while' and 'whilst'. In the 14 essays known to be by Madison, 'while' never occurs; 'whilst' occurs in eight of them. 'While' occurs in 14 of 48 Hamilton essays, 'whilst' never. Thus the presence of 'whilst' in five of the disputed papers pointed towards Madison as author.

However, such markers were few in number and Mosteller and Wallace found it useful to look for words which were used by both authors with comparative frequency, but at different rates. Words such as prepositions, conjunctions, articles—'function words' whose frequency is almost independent of context—provided suitable instances. It was found, for instance, that low rates for 'by' were characteristic of Hamilton, and high rates of Madison; with 'to' the situation was reversed. After considering several thousand function and other words, Mosteller and Wallace found 28 with strong discriminating power: the set included 'upon', 'also', 'an', 'by', 'of', 'on', 'there', 'this', 'to' among function words and 'innovation', 'language' and 'probability' among less context-free words. On the basis of comparisons between the rates of use of these words, Mosteller and Wallace felt able to conclude beyond reasonable doubt that Madison wrote the 12 disputed papers.

Mosteller and Wallace used far more sophisticated statistical methods than Ellegård had done, though their discrimination was based on a smaller set of stylistic indicators. The methods they used are more elaborate than any that will be described in this book. Indeed, they were themselves more interested in testing methods of statistical discrimination than they were in the authorship attribution as an end in itself. Their methods were based on a statistical result known as Bayes' theorem which enables initially assessed odds on an outcome to be constantly revised as fresh evidence is adduced. The use of Bayes' theorem enabled them to calculate the final odds on the authorship of each essay. The odds in favour of Madison were astronomical in the case of all essays except two, and even there the odds were 800 to 1 and 80 to 1. Like Ellegård, Mosteller and Wallace were careful to combine historical evidence with statistical computation, and their book *Inference and Disputed Authorship: The Federalist* has been widely admired. It contains much of general value in literary statistics: but it is difficult reading for the novice.

An ancient puzzle which in some ways stands between the Junius and the Federalist problems is presented by the Aristotelian corpus. This contains two ethical treatises parallel in structure: the *Nicomachean Ethics* and the *Eudemian Ethics*. The relationship between them is not a problem of authorship attribution: scholars nowadays agree in attributing both works to Aristotle. The puzzle is that three books make a double appearance in the manuscript tradition of the treatises: once as books five, six and seven of the *Nicomachean Ethics*, and once as books four, five and six of the *Eudemian Ethics*. In 1978 I published a study (*The Aristotelian Ethics*) in which I argued that stylometric indications suggested overwhelmingly that the disputed books resembled the Eudemian context rather than the rival Nicomachean one.

The problem resembled the Federalist rather than the Junius one in being a simple matter of assigning a disputed text to one of two competing contexts; it was a question of deciding whether text A resembled text B more than it resembled text C. But the number of features which turned out to discriminate the *Nicomachean Ethics* from the *Eudemian Ethics* was much larger than the number of discriminants between Madison and Hamilton. Significant differences were found in particle use, in the use of prepositions and adverbs, pronouns and demonstratives, in the preference for different forms of the definite article. Altogether 24 independent tests were carried out based on some 60% of the total word-usage in the text, and in 23 out of 24 cases the tests gave an unambiguous answer that the common books, considered as a whole or in small samples, resembled the Eudemian treatise more than the Nicomachean one. The tests used were largely inspired by Ellegard's methods; the wealth of the material and the straightforward nature of the problem meant that simple statistical methods were adequate to establish the conclusion with a high degree of probability.

In the course of investigating differences between the two ethical treatises I observed that a group of adverbs and expressions indicating tentativeness—words for 'perhaps', 'it seems likely', etc.—were more than twice as common in the Nicomachean books as in the Eudemian. By contrast, words expressive of confidence—'certainly', 'obviously' and 'clearly'—are about three times as popular in the *Eudemian Ethics* as in the Nicomachean. It is not clear what, if any, significance should be attached to this in the Aristotelian context; but a systematic study of such expressions of certainty and hesitation has been made by the German sociologist Suitbert Ertel. Ertel has identified six categories of words within which authors have a choice between dogmatic and tentative alternatives. In the category of frequency, for instance, there are the dogmatic words 'always', 'whenever', 'never' and so on, contrasted with the less confident 'often', 'sometimes', 'occasionally'. In the category of degree there are adverbs like 'absolutely', 'unreservedly' to contrast with 'up to a point', 'to a certain degree' and so on. On the basis of the words selected from one or other set of alternatives within each category, Ertel assigns writers a 'Dogmatism quotient'. His results, though not perhaps surprising, are interesting. Among philosophers it turns out that Marx and Heidegger have a high dogmatism quotient, while Russell and Locke have a low one. Of the speeches made in the Reichstag during the Weimar Republic, those of the Communists and National Socialists score high on the dogmatism scale: the low scorers are the Social Democrats and other politicians of the centre. The dogmatism-quotients of individuals can be studied over time: Ertel shows how Hitler's quotient rose in moments of crisis and sank in times of victory. Immanuel Kant's quotient mounted during his precritical period, rose to a climax with the second edition of the *Critique of Pure Reason*, and then gently declined with advancing age.

As Ertel's work illustrates, the use of statistical techniques in literary studies is by no means confined to the area of authorship attribution or the determination of chronology. Many statistical studies have been made of features of verse such as metre, rhythm, alliteration, harshness and softness of sound. Such studies have been made in many languages: there are, for instance, the studies of Homer by David Packard, of Virgil by L. Ott, of Corneille by Ch. Muller, of Shakespeare and Swinburne by B. F. Skinner. Here as elsewhere contemporary workers are following, with more sophisticated statistical apparatus, routes first mapped at the end of the last century. To this day scholars defer to the dating of Euripides' plays carried out in 1906 by Zylinski by the increasing number of resolved feet in his verses.

One important recent development in stylometry should be mentioned. From 1974 A. Q. Morton and his associates turned their attention from Greek to English. In recent years thay have studied disputed Shakespeare plays, modern imitations of Jane Austen and of Conan Doyle, and a number of other literary problems. They have made use of Ellegård's methods of analysing synonym choices, but also developed powerful new techniques by studying the position and immediate context of word-occurrences. Such features as the proportion of occurrences of 'the' preceded by 'of', or the number of times 'and' occurs as the first word of the sentence, provide, it is claimed, indicators of authorship which can be used to settle the attribution of even quite short samples of text. These techniques have even been presented before courts as a method of discriminating between genuine and fabricated confessions to the police. In at least one case evidence of this kind has led to the acquittal of the accused on all charges in which the disputed statements were cited. (See Morton, *Literary Detection*, 206.)

The U.S. courts have not, so far, followed the precedent thus set in English courts. When Patricia Hearst was tried for bank robbery an important part of the evidence consisted of tape-recorded revolutionary statements spoken by members of the Symbionese Liberation Army. Hearst's counsel, F. Lee Bailey, challenged the attribution of these statements to the accused, and sought leave to introduce expert 'psycholinguistic' testimony on the issue of authorship. The trial judge ruled against admitting the testimony, saying that an aura of special reliability might have attached to the expert's testimony which would have been unjustified in view of 'the relative infancy of this area of scientific endeavour'.

Even among those most familiar with statistical authorship studies there were many who shared the misgivings of the judge about using statistical stylometry, at the present stage of its development, for forensic purposes. Professor Richard Bailey of the University of

Michigan, a doyen of the study of statistics and style in the U.S., in a most useful paper outlines the difficulties facing the would-be expert witness, and says:

> In my view, there are at least three rules that define the circumstances necessary for forensic authorship attribution.
> 1. that the number of putative authors constitute a well-defined set.
> 2. that there be a sufficient quantity of attested and disputed samples to reflect the linguistic habits of each candidate.
> 3. that the compared texts be commensurable.

The third point is the one which remains the most serious difficulty in courtroom cases: differences between letters, diaries and oral confessions by the same person are likely to be much greater than those between two persons writing in the same mode of composition. Bailey's paper continues with a detailed consideration of the material which was disallowed at the Hearst trial and concludes that had expert testimony been admitted it is unlikely that it would have been unequivocally favourable to the accused.[1]

No doubt there will be future occasions on which the accused in criminal trials will seek to use stylometry to challenge confessions attributed to them, or in which the litigants in civil actions will use similar methods of textual study to contest the authenticity of contracts and wills. However the discipline may develop in the future it cannot be said that there is as yet such a thing as a stylometric fingerprint: a method of individual style which is as reliable as a fingerprint as a criterion of personal identification.

What would a stylistic fingerprint be? It would be a feature of an author's style—a combination perhaps of very humble features such as the frequency of *such as*—no less unique to him than a bodily fingerprint is. Being a trivial and humble feature of style would be no objection to its use for identification purposes: the whorls and loops at the ends of our fingers are not valuable or striking parts of our bodily appearance. But the item would have to be a constant feature of an author's writing, as fingerprints remain the same throughout life, and it would have to be unique to him and shared by no other writer. At the present time no one knows whether there are such

---

[1] 'Authorship attribution in a forensic setting', in *Advances in Computer-aided Literary and Linguistic Research*, ed. Ager, Knowles & Smith.

features of style as not enough data have been collected. Constancy is not too difficult to test and may in some cases have been proved: uniqueness is quite another matter.

At present it seems possible to identify statistical features of style, objectively measurable, which are unique to particular authors in the sense that they appear in all the writings of that author and not in any other so far studied. But the number of authors in any language who have been studied in this way is only a fraction of those who remain to be studied. The effects on stylistic habits of individual authors of variations of style and genre has not yet been widely studied and cannot be said to be well understood.

Some students believe that, given time, a body of material will be accumulated which will permit the positive identification of every individual from a small sample of his writing. After all, they can point out, a considerable time elapsed between Galton hypothesizing the uniqueness of fingerprints and the adoption of the fingerprint for identification purposes in 1901. Perhaps—some say—by 2001 stylometric profiles of all citizens will be on file in the F.B.I. if not yet in Scotland Yard.

The hope—or fear—is not total fantasy. As the art of stylometry progresses it may be possible to identify features of the style of a long-dead author which are so uniform throughout his writings, and so peculiar to him in comparison with contemporaries writing in the same style and genre, that they can be used with virtual certainty to settle whether a document putatively attributed to him is or is not genuine. Literary stylometry may develop into a genuinely scientific discipline with a high degree of reliability.

But what is true of long-dead authors will not necessarily be true of contemporary authors. The methods in question can only be applied with anything approaching certainty to documents prior to the discovery of stylometric techniques. One cannot alter one's fingerprints or forge another person's but once a feature of style has been identified, it becomes subject to choice—whether of an author himself or of his imitators. To write a piece of prose answering to a set sentence-length distribution, or to copy given habits of collocation and synonym choice is not a trivial task. But, given time and motive, it can be done.

So there can be no definitive test of authorship: only a constant game of pat-a-cake between forgery and detection, as in the imitation of works of art as fashions change.

Stylometry can hope in time to fulfil the aspirations of those who take it up, as one tool among many, for literary and historical research. It cannot fulfil the hopes or fears of those who see it as an extension to the mental sphere of the individuating techniques we use in the physical identification of bodies.

But it is not the intention of the present work to evaluate, or even to survey, the claims advanced and the results recorded by those who have used statistical methods in the study of literary texts. Rather, it is to instruct readers who have a humanistic background in the elementary knowledge of statistical technique which will enable them to assess for themselves the uneven contributions of scholars in this field. To this task the rest of the book is devoted. The purpose of the present chapter has been simply to illustrate the types of task in which the scholar can employ the statistical techniques now to be explained.

2

Distributions and Graphs

IF WE are to make a mathematical study of an author's style we must first identify features of his writing which can be precisely quantified: countable events and measurable magnitudes. Such features are not difficult to find. The length of words and sentences and paragraphs can be measured; it is possible to count the frequency of vocabulary items or syntactic constructions or rhetorical devices. By measuring and counting stylistic traits of such kinds we can hope to discover regularities characteristic of a particular author or a particular genre. Elementary statistical theory will enable us to give compact descriptions of the quantifiable features of style; it will also tell us what degree of confidence to place in generalizations about the style of an author based on a small sample of his work. It will assist us to make systematic comparisons between the stylistic habits of different writers, and permit us to distinguish genuine differences of style from chance variations in usage to which we should attribute no more importance than we would to a particular lucky or unlucky hand dealt at bridge. In this and other ways the statistical study of style may help to make more precise the analysis of stylistic peculiarities of writers and may contribute to the solution of traditional problems concerning the authenticity, integrity and chronology of literary works of uncertain dating or provenance.

To introduce the statistical study of style we may begin with a simple and readily observable feature of any piece of writing: the length of the words which occur in it, counted in letters. Consider the following passage from Jane Austen's *Northanger Abbey.*

> She entered the rooms on Thursday evening with feelings very different from what had attended her thither the Monday before. She had then been exulting in her

15

engagement to Thorpe, and was now chiefly anxious to avoid his sight, lest he should engage her again; for though she could not, dared not expect that Mr Tilney should ask her a third time to dance, her wishes, hopes and plans all centred in nothing less. Every young lady may feel for my heroine in this critical moment, for every young lady has at some time or other known the same agitation.

It is well within the mathematical powers of the most innumerate of us to count the number of letters in each successive word in this passage and to record the result. If we do so we may obtain the following list of numbers:

3	7	3	5	2	8	7	4	8	4	9	4	4	3	8	3	7	3
6	6	3	3	4	4	8	2	3	10	2	6	3	3	3	7	7	2
5	3	5	4	2	6	6	3	5	3	6	3	5	3	5	3	6	4
2	6	6	3	3	1	5	4	2	5	3	6	5	3	5	3	7	2
7	4	5	5	4	3	4	3	2	7	2	4	8	6	3	5	5	4
3	2	4	4	2	5	5	3	4	9								

Even this first step in quantification is not wholly straightforward. 'Mr' is an abbreviation for 'Mister': should it count as two letters or six? Arbitrarily, we count it as two letters. In quantifying stylistic features we are constantly faced with the need for similar decisions: commonly, what matters is not which decision is taken, but that an arbitrary decision, once taken, should be consistently adhered to.

The more serious problem is this: what is the use of such a list of numbers once we have it? The answer is that as it stands it is almost wholly useless, but it can be rearranged in such a way as to bring out clearly the information which it contains. For the great majority of purposes data of the kind illustrated in the table above are best presented in what is known as a *frequency distribution*.

To construct a frequency distribution of a set of measurements we must first make a list of classes into which the measurements fall. In the present instance every word occurring in the passage is between one and ten letters long, so the obvious classes to choose are the ten classes: one-letter words, two-letter words, three-letter words, and so on up to ten-letter words. We then arrange these classes in a column, with the class of ten-letter words at the top and the class of one-letter words at the bottom, and put the column on the left-hand side of the page. We then go through our list of measurements and each time we meet a three-letter word we put a tick against the three-letter class; each time we meet a four-letter word we put a tick against the four-letter class; and in general each time we meet an X-letter word we put a tick against the X-letter class. It helps to keep these scores tidy and perspicuous if every fifth tick is made across the previous four ticks: this presents us with the score in convenient packets of five units which makes it easier to calculate the total. When we have completed the list of measurements we total the scores in each class and write the totals in a column on the right-hand side of the page in the appropriate position corresponding to the place where the class was listed on the left-hand side. If we do this with the passage from Jane Austen we obtain a result looking something like this:

TABLE 1.1. *Length in letters of words in a 100-word passage of* Northanger Abbey

Length of words in letters Class	Tally of scores	Number of such words Frequency
10	I	1
9	II	2
8	︴HHr	5
7	︴HHr III	8
6	︴HHr ︴HHr I	11
5	︴HHr ︴HHr ︴HHr I	16
4	︴HHr ︴HHr ︴HHr II	17
3	︴HHr ︴HHr ︴HHr ︴HHr ︴HHr II	27
2	︴HHr ︴HHr II	12
1	I	1
Total		100

This table gives us a frequency distribution of word-lengths for the passage. The essential part of the table is the column of class labels on the left and the column of frequencies on the right: the score tallies are shown to illustrate how to construct such a table, but published frequency distributions naturally omit this item of machinery and simply print the two columns of numbers. Again, in constructing a frequency distribution such as the above, it is not necessary first to list the measurements in undistributed form as we did above: we can simply go through the passage word by word and make a tick every time in the appropriate class. Finally, as above, we add the figures in the frequency column to obtain the total number of cases, in this case the total number of words in the passage.

Many things are easily seen in the frequency distribution which were not obvious in the original passage and were hidden in the undistributed list of measurements. Thus we can see that three-letter words are more frequent in the passage than words of any other particular length; we can see that the majority of words which make up the passage are between two and six letters long; we can see that if we consider only words of three or more letters, the longer a word is the less frequently words of that length occur in the passage, and so on. On the other hand, we have lost some of the information which was present in the original passage and in the list of the lengths of successive words. Suppose, for instance, that we were interested in the occurrence of long and short words, defined respectively, in an arbitrary manner, as words with at least five letters and words of less than five letters. If we wanted to know whether short words were more frequently followed by long words or by short words, we could find this out by inspecting the undistributed list, since this preserved the order of words; but the frequency distribution will tell us nothing about it. We often find, as in this case, that we can make certain items of information more perspicuous only by throwing away other items.

Considered in itself, the frequency distribution we have just constructed contains information only about one small passage of *Northanger Abbey*. It does not contain data about any other of Jane Austen's works, or about the English novel in general, or about early 19th-century English prose. We may wonder how far it is representative of the various categories of writing to which it belongs. If we took any other 100-word sample of Jane Austen's writing would we find that the three-letter words were the commonest? Would we, indeed, find that each 100-word sample contained exactly 27 three-letter words, so that three-letter words made up exactly 27% of her writing? Would we perhaps even find that the popularity of the three-letter word was a feature of the English language in general, so that any passage of comparable length would throw up more three-letter words than words of any other size?

A full answer to these questions would have to fall into two parts. The first part would consist of mathematical theory concerning the relationship between frequency distributions for small quantities of data and frequency distributions for larger bodies of data from which these small quantities are drawn. It will be the purpose of later chapters to give an elementary presentation of this theory. The second part of the answer would be not theoretical but empirical: it would consist of information concerning the regularities to be observed by the actual study of English writers. A considerable amount of such information is now available. This book will not be directly concerned in presenting it, though it will indicate some of the places where it is to be found. But the reader may like to try his hand at collecting a scrap or two of such information by working through the following exercise, which is designed to familiarize him with the construction of a simple frequency distribution.

Exercise 1

Construct a frequency distribution of the length of words in letters for the following 100-word passages of English prose.

(a) Only imagine a man acting for one single day on the supposition that all his neighbours believe all that they profess, and act up to all that they believe. Imagine a man acting on the supposition that he may safely offer the deadliest injuries and insults to everybody who says that revenge is sinful; or that he may safely entrust all his property without security to any person who says that it is wrong to steal. Such a character would be too absurd for the wildest farce. Yet the folly of James did not stop short of this incredible extent. (Macaulay, *Essay on Sir James Mackintosh.*)

(b) But there is a point of depression as well as of exaltation, from which human affairs naturally return in a contrary direction, and beyond which they seldom pass, either in their advancement or decline. The period in which the people of Christendom were the lowest sunk in ignorance, and consequently in disorders of every kind, may justly be fixed at the eleventh century, about the age of William the Conqueror; and from that era the sun of science, beginning to reascend, threw out many gleams of light, which preceded the full morning when letters were revived in the fifteenth century. (Hume, *History of England*, Chapter 23.)

Solution to Exercise 1

(a) Passage from Macaulay

Length of words in letters	Frequency
11	2
10	2
9	3
8	3
7	10
6	9
5	8
4	15
3	29
2	16
1	3
Total	100

(b) Passage from Hume

Length of words in letters	Frequency
12	1
11	2
10	2
9	7
8	4
7	9
6	9
5	13
4	13
3	18
2	20
1	2
Total	100

The information contained in a frequency distribution may be presented more clearly by being laid out in graphical form. Any reader of newspapers is familiar with graphs: they are two-dimensional representations of sets of data. To construct a graph we must first draw two lines perpendicular to each other. These lines are called axes, and a scale is associated with each, the scales being usually indicated by tick marks and labels. A pair of values which can be separately indicated by one point on each scale is represented by the intersection of lines perpendicular to the corresponding axis. The graphs which we are most familiar with in everyday life are time series graphs: the horizontal axis is divided into a time-scale (of years, or months, or days, for instance) and the vertical line shows the magnitude of some measurable phenomenon at the times indicated by points along the horizontal scale. (A barograph, for instance, will record in a graph the atmospheric pressure at various times during the day.) But the horizontal line, the X-axis, as it is often called, need not represent time: it may represent any feature which can occur with various magnitudes or take on various values. When we use a graph to represent a frequency distribution the values which appear along the horizontal scale are the values which identify the different classes of the distribution. To represent the information given above, we scale the horizontal line by marking along it the various numbers which may replace the variable letter X in the description 'word of X letters'. The vertical line (or Y-axis) is scaled so as to represent the number of times which members of particular classes may occur: in the case in point, the number of times which words of a particular length are found. The letter f for frequency is often used to mark this variable.

Suppose, for instance, that we wished to represent graphically the lengths of the words in the following passage:

Fortunately it happens, that since reason is incapable of dispelling these clouds, nature herself suffices to that purpose, and cures me of this philosophical melancholy and delirium, either by relaxing this bent of mind, or by some avocation, and lively impression of my senses, which obliterate all these chimeras. I dine, I play a game of backgammon, I converse and am merry with my friends; and when after three or four hours' amusement, I would return to these speculations, they appear so cold, and strain'd and ridiculous, that I cannot find in my heart to enter into them any farther. (Hume, *Treatise of Human Nature*, I, IV.)

We might construct a frequency distribution as before and on its basis draw the following graph.

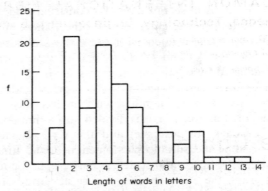

Figure 1. Graph of word-lengths in a 100-word passage of Hume.

Such a graph is called a *histogram*. In a graph of this type if we allow equal parts of the X-axis to represent classes of equal width, then the frequency of each kind of item is represented by an area in the form of a vertical bar whose base is that part of the X-axis marked off to represent the class to which items of that kind belong. The height of the bar is determined by the point on the vertical scale corresponding to the appropriate frequency. Thus, as there are six one-letter words in the passage of Hume, the bar above the part of the X-axis marked '1' is six units high. The empty space over the section of the X-axis marked '14' represents that there are no 14-letter words in the passage. Since the intervals along the X-axis are all equal the area of each bar, as well as its height, will correspond to the frequencies associated with the various classes. Obviously, the frequencies could be represented by lines instead of by rectangles, but the representation by rectangles has a more satisfactory visual impact, and this is why histograms are popular for the graphical representation of frequency distributions.

Exercise 2

(a) Write out the frequency distribution represented pictorially in the histogram in Figure 1.

(b) Construct histograms from the frequency distributions for the passages (a) and (b) in Exercise 1.

Solution to Exercise 2.

(a)

Length of words in letters	Number of such words
13	1
12	1
11	2
10	5
9	3
8	5
7	6
6	9
5	13
4	19
3	9
2	21
1	6
	100

(b)

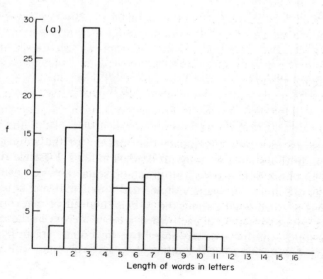

Length of words in letters

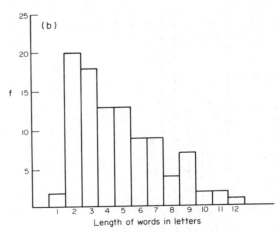

With the aid of squared paper, a histogram may be constructed while the data for a frequency distribution are being assembled.

Thus consider the first verse of a well-known song by Shakespeare:

> Blow, blow, thou winter wind
> Thou art not so unkind
>   As man's ingratitude
> Thy tooth is not so keen
> Because thou art not seen,
>   Although thy breath be rude.
> Heigh ho! sing, heigh ho! unto the green holly:
> Most friendship is feigning, most loving mere folly:
>   Then heigh ho, the holly!
>   This life is most jolly.

If we wish to construct a frequency distribution for the word-lengths occurring in this stanza we may proceed as follows. First we rule the vertical and horizontal axes of our graph on squared paper, using one square as the unit of frequency along the Y-axis and one square for each class of word-length along the X-axis. Instead of marking ticks when we count the letters of each word, we put a cross in the column corresponding to the number given by the count, starting at the bottom and working upwards with each new cross. The result will be

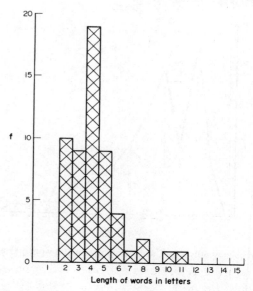

Figure 2. Histogram of the first verse of *Blow, Blow thou Winter Wind*, word lengths in letters.

something like Figure 2: to make a histogram we merely have to rule a line round the areas marked by the 'x's we have filled in. We can then read off the frequency distribution from the histogram.

As will be seen, the use of a single square for the unit along each axis produces a rather ungainly histogram: applied to a passage longer than 50-odd words the method would produce a graph with an even uglier vertical elongation. When graphing a distribution, it is best to aim at having the tallest bar approximately three-quarters as high as the horizontal axis is long. This can be achieved by appropriate choice of unit-lengths along each scale.

A histogram is not the only method of representing a frequency distribution in graphical form. Another is the *frequency polygon*. To set up a frequency polygon the vertical axis of the graph is marked off in units of frequency as before; but along the horizontal axis, instead of marking off intervals of equal length to correspond to the classes, we mark off points at equal distances to represent the classes. We

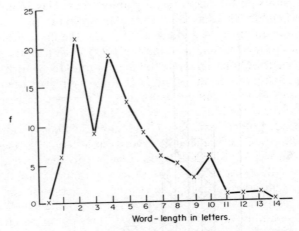

Figure 3. Frequency polygon of word-lengths in a 100-word passage of Hume.

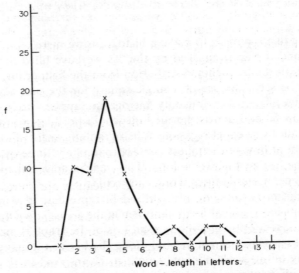

Figure 4. Frequency polygon of word-length in a song of Shakespeare.

then plot points on the graph directly above these class points on the X-axis, at heights corresponding to the appropriate frequency. If we connect these points by straight lines, and join the first and last point of the series to the baseline, we obtain a polygon which represents the frequency distribution. Thus the data of Table 1.1, which are represented in histogram form in Figure 1, could be represented in polygon form in Figure 3; and corresponding to the histogram of Figure 2, we could draw the polygon of Figure 4.

A frequency polygon is a less natural method than a histogram to represent data such as the length of words in letters. The lines joining the points in Figures 3 and 4 are merely aids to the eye and do not represent anything about the data in the way that the points do. This is clear enough: a word cannot consist of $3\frac{1}{2}$ letters, nor can it occur $7\frac{1}{3}$ times. The length of a word in letters is a variable which can take only whole-number values: it is, in technical terms, a discrete and not a continuous variable. If we wished to represent graphically data about a continuous process such as the growth of children, the frequency polygon is much more appropriate. The points plotted on the graph may represent the height of a child on each of its birthdays; but as the child continues to grow between one birthday and the next, the suggestion of continuity made by the line joining the plotted points is not inappropriate, and the line may even help us to estimate the height at a date between birthdays.

Provided we are prepared to discount the misleading suggestions of continuity in a frequency polygon, for several purposes we may find it a more convenient graph to use than a histogram. If we wish, for instance, to represent two different distributions on the same graph, with a view to comparing them, we can do so much more perspicuously by plotting two frequency polygons than by superimposing two histograms. In Figure 5 the frequency distributions for the word-lengths of two passages from Hume and Macaulay are represented by two frequency polygons on a single graph.

Various properties of frequency distributions can be very conveniently indicated by describing the shape of the frequency polygon which corresponds to them. A polygon is *symmetrical* if one side of it is the mirror image of the other: a distribution, likewise, is called symmetrical if its frequency polygon is symmetrical, as in Figure

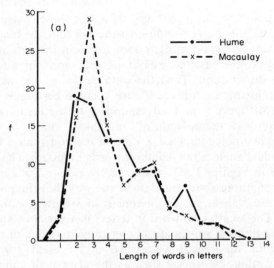

Figure 5. Frequency polygons for word-length distributions in passages from Hume and Macaulay.

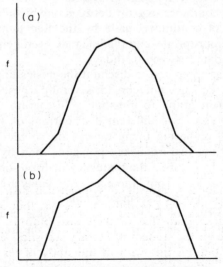

Figure 6. Symmetrical frequency polygons.

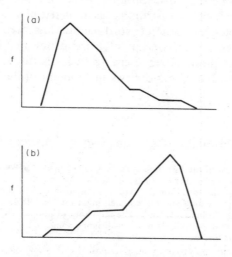

Figure 7. Skewness: (a) positive; (b) negative.

6(a and b). Most frequency distributions—including all the ones we have so far looked at—are not symmetrical, but extend farther in one direction than in the other, with their polygons exhibiting a 'tail'. Such distributions are called *skew* or *skewed*, as in Figure 7. They are called *positively skew* if the tail is to the right, towards the higher values on the *X*-axis, and *negatively skew* if the tail is to the left, towards the lower values. The polygon in Figure 7(a) represents a positively skew distribution, and that in Figure 7(b) represents a negatively skew one.

The mathematical treatment of symmetrical distributions, and in particular of bell-shaped distributions such as that in Figure 7(a), is considerably easier than that of skew distributions. This fact is an unfortunate one for the student of literary statistics, since most of the frequency distributions met with in the study of texts are definitely skew. But the difficulty is not an insuperable one, as there are methods, which we shall indicate in later chapters, for enabling the theory of symmetrical distributions to be brought to bear on the study of data which are distributed asymmetrically.

In this chapter the topic of word-lengths has been used to introduce the reader to the concept of a frequency distribution and to illustrate the methods of representing such distributions numerically and graphically. Exercise 3 reviews the content of the chapter.

Exercise 3

Take the following passage from Sydney's *Apology for Poetry*:

So, then, the best of the historian is subject to the poet; for whatsoever action or faction, whatsoever counsel, policy or war-stratagem the historian is bound to recite, that may the poet, if he list, with his imitation make his own, beautifying it both for further teaching and more delighting, as it pleaseth him; having all, from Dante's Heaven to his Hell, under the authority of his pen. Which if I be asked what poets have done so? as I might well name some, yet say I, and say again, I speak of the art and not of the artificer.

(a) Construct a frequency distribution for the length in letters of the words occurring in this passage.

(b) By constructing a histogram, or a frequency polygon for the frequency distribution for the passage, decide whether the distribution is symmetrical, positively skew or negatively skew. Which of the other distributions studied in this chapter does it resemble in this respect?

Solution to Exercise 3.

(a)

Length of word in letters	Number of such words
12	1
11	1
10	4
9	4
8	2
7	5
6	5
5	8
4	18
3	27
2	21
1	4
	100

(b) Positively skew, as are all the others.

3

Measures of Central Tendency

A HISTOGRAM or a frequency polygon gives a detailed and perspicuous representation of a frequency distribution. However, if we wish to record and compare a large number of frequency distributions, it is cumbrous and uneconomical to produce a graph of each single one. We need a concise method of expressing the salient features of a distribution. The two most important things to know about a frequency polygon is where it is centred and how it is spread out. Statisticians have devised measures which will give this information in compact numerical form. Measures which indicate the central location of a distribution are called *measures of central tendency*. Measures which indicate the way in which a distribution is spread out are called *measures of variability* or *measures of dispersion*. The present chapter will be concerned with measures of central tendency, and the succeeding one with measures of dispersion.

The everyday word for a measure of central tendency is 'average'. Everyone has some idea what an average is, but there are in fact several different kinds of average and the popular notion combines in a muddled manner features characteristic of the different kinds. In this chapter we will describe three distinct measures of central tendency: the mode, the mean and the median.

The mode is the simplest and crudest index of central tendency: it is simply the peak, or highest point, of the frequency polygon. A glance at Figure 1 suffices to show that the mode of word-length in the passage from Hume's *Treatise* is two letters. In the passage from Hume's *History* in Exercise 1(b) the mode again occurred at the two-letter point, whereas in the passages we have studied from Jane Austen and Macaulay the modal length was three letters, and in the

Shakespeare song it was four letters. In giving the mode of a distribution you are merely indicating the most popular value on the X-axis, the class with the most members.

It will be recalled from Chapter 1 that some of the earliest recorded studies of literary statistics concerned the mode of the word-length distribution. Of the authors studied by T. C. Mendenhall almost all peaked at the three-letter word: they included Dickens, Thackeray and a number of Shakespeare's contemporaries. Only Marlowe shared Shakespeare's preference for the four-letter words; the only other author studied by Mendenhall to depart from the three-letter mode was John Stuart Mill, who peaked at the two-letter word.

The mode, though the most easily calculated, is the least useful of the measures of central tendency. It cannot be made the basis of any further statistical calculation and it may give a misleading impression of where the majority of data in a distribution lie. Moreover, a distribution may not have one single mode: there may be no single value which occurs more frequently than any other. An author, for instance, may use three- and five-letter words with equal frequency. In such a case the word-length distribution would be said to be *bimodal*, as having two modes. The frequency polygon for a bimodal distribution will resemble a mountain range dominated by two high peaks.

Exercise 4

Divide the following passage into three sections of equal length in words. By constructing a frequency distribution for each section, determine the mode of word-length in each case and the frequency of the modal value.

How does the objector know that women do not desire equality and freedom? He never knew a woman who did not, or would not, desire it for herself individually. It would be very simple to suppose, that if they do desire it they will say so. Their position is like that of the tenants or labourers who vote against their own political interests to please their landlords or employers; with the unique addition, that submission is inculcated on them from childhood, as the peculiar attraction and grace of their character. They are taught to think that to repel actively even an admitted injustice done to themselves, is somewhat unfeminine, and had better be left to some male friend or protector. To be accused of rebelling against anything which admits of being called an ordinance of society, they are taught to regard as an imputation of a serious offence, to say the least, against the proprieties of their sex. It requires unusual moral courage as well as disinterestedness in a woman, to express opinions favourable to women's enfranchisement until at least there is some prospect of obtaining it. The comfort of her individual life, and her social consideration, usually depend on the good will of those who hold the undue power.... (Mill, *The Enfranchisement of Women*.)

Answer to Exercise 4.

Length of word in letters	Number such I	II	III
17	0	0	1
16	0	0	0
15	0	0	1
14	0	0	0
13	0	0	1
12	1	0	0
11	0	0	1
10	0	5	3
9	5	6	1
8	3	6	3
7	5	3	8
6	5	7	4
5	8	6	10
4	13	11	6
3	13	7	11
2	16	19	18
1	1	0	2
	70	70	70
Mode	2	2	2

A commoner and more valuable measure of central tendency than the mode is the *mean*. It is the mean (strictly speaking 'the arithmetic mean') which comes nearest to corresponding to the layman's conception of an average. To calculate the mean you add the values of all the data in the distribution and then divide this sum by the total number of data. Thus, suppose that we wish to obtain the mean word length in the following sentence:

Now is the time for all good men to come to the aid of the party.

We add together the numbers which give the length of each word, thus:

$$3 + 2 + 3 + 4 + 3 + 3 + 4 + 3$$

$$+ 2 + 4 + 2 + 3 + 3 + 2 + 3 + 5$$

This gives us a total of 49 letters. We divide this sum total by the number of words, which is 16, and we get the result $3\frac{1}{16}$ or 3.0625.

Whereas the mode of word-length can only be a whole number, the mean word length, as in this instance, may take fractional values. In general, the mode can only take the same sort of values as the data in the distribution; the mean may take any value whether the data can take only whole-number values or not. It is in this sense of average that we can say that the average word contains 3.0625 letters, though no actual word could contain 3.0625 letters.

The mathematical formula for the calculation of the mean is

$$\mu = \frac{\Sigma X}{N}$$

Σ is the capital Greek letter sigma, which tells us to sum, or add, what follows; X is a variable letter for the values in the distribution, and N is the number of items in the distribution; μ, the Greek letter mu, is the symbol for the mean. (This is used for the mean of a whole population; if we take samples from a population we denote the sample mean by $\bar{X}$.)

The mean may be calculated from a frequency distribution like those given in the previous chapter. Where particular values (e.g. word-lengths) occur more than once, each value must be multiplied by its frequency before being summed. To illustrate this, let us consider the frequency distribution for the first 70 words of the passage of Mill in the previous exercise.

Word length	Number such	Length × frequency
X	f	fX
12	1	12
11	0	0
10	0	0
9	5	45
8	3	24
7	5	35
6	5	30
5	8	40
4	13	52
3	13	39
2	16	32
1	1	1
	70	310

$\mu = \Sigma f X / N = 310/70 = 4\tfrac{3}{7} = 4.429$.

The mean length of words in the passage is the sum of all the products obtained by multiplying length by frequency, divided by the total number of words. (This will, of course, be the same as the result of dividing the total number of letters in the passage by the total number of words.) In this case it is 310 divided by 70, which is $4\frac{3}{7}$ or 4.429. The means of the other two passages can be calculated as 4.971 and 4.914.

Exercise 5.

Calculate the mean word-lengths of the passages from Hume and Macaulay in Exercise 1(b).

Solutions: Macaulay 3.85; Hume 4.97.

Most people, if asked to work out the average of a set of values, will work out, in roughly the manner we have described, the mean. However, there are many uses of 'average' in ordinary language in which it refers not to the mean but to some different measure of central tendency. For instance, if a newspaper talks about the income of the average man, it probably means how much is earned by a man who is richer than half the rest of the population and poorer than half the rest of the population. What we are looking for here is not the mean income but the *median* income. The median is the middle point in a distribution, the point which divides it into two equal halves.

Mean and median may differ quite spectacularly. A community consisting of one millionaire plus his staff of four, earning yearly respectively £10,000, £8,000, £6,000 and £4,000 will have a median annual income of £8,000, since the man earning that sum is richer than two others, and poorer than two others, of the total population. But its mean annual income is: (£1,000,000 + £10,000 + £8,000 + £6,000 + £4,000)/5, that is to say, £205,600. As this example shows the mean, unlike the median, is affected very much by a single extreme value. A single exceptionally long word in a passage ('disinterestedness' in the Mill passage, for instance) will raise the mean word-length while not affecting the median word-length.

If there is an odd-number of values in a distribution, and if each value occurs no more than once, as in the artificially simple example of the millionaire's household, the median can be discovered by simple inspection if the values are placed in ascending or descending order. In the great majority of cases where the values are set out in a frequency distribution the median requires calculation. The first step in this calculation is to convert the frequency distribution into a *cumulative frequency distribution.*

A cumulative frequency distribution is a distribution recorded in such a way as to indicate how many of the items in the distribution fall below particular values. A cumulative frequency distribution for word-length, for instance, will show how many words in a text are one letter long, are two letters long or less, are three letters long or less, and so on. To produce a cumulative frequency distribution you simply take the frequency distribution and add, from the bottom upwards, the number of items in each class. Thus, we can turn the Mill distribution in Table 2.1 into a cumulative frequency distribution thus:

Word-length	f	cf
12	1	70
11	0	69
10	0	69
9	4	69
8	4	65
7	6	61
6	4	55
5	8	51
4	13	43
3	13	30
2	16	17
1	1	1

From this table we can see, for instance, that 30 words out of the total 70 have a length of three letters or less, and that 69 words are of 10 letters or less.

To discover the median word length we need to discover which 50% of the words in the text fall below and which 50% fall above. We can say that the median value is the length of that word which has equal numbers to right and left of it when the words are arranged in length order. This would suggest the value 4. But there is no whole-number value such that 35 words, or 50% of the words, have a length equal to or less than that value, so that we have to interpolate a value between two whole numbers for the median. The median, like the mean and unlike the mode, can take fractional as well as integer values.

To interpolate a median we have to pretend that word length is a continuous rather than a discrete quantity, so that a word may have a length like 3.456 letters. We will go on to assume on this basis that our three-letter words are words of any length from 2.5 to 3.5 letters, and our four-letter words are words of any length from 3.5 to 4.5 and so on.

We know from the cumulative frequency distribution that the median word length must come somewhere between the upper limit for three-letter words, and somewhere below the upper limit for four-letter words. For less than 50% of the words in the text (namely 30, or approximately 43%) have a length of three letters or less, and more than 50% of the text (43, or approximately 61%) fall in the category of words classed as four letters or less.

To reach 50%, or 35 words, we need to take five of the four-letter words in addition to the 30 words of three letters or less. As there are 13 four-letter words, we are taking $\frac{5}{13}$, or 0.38, of them to make a total of 35 words. To obtain the median, we add this 0.38 to the upper limit for three-letter words, namely 3.5, and obtain 3.88 as the median word length of the passage. (It might seem more natural to add the 0.38 to the value 3, rather than to 3.5; but since, in order to interpolate a median we are making the assumption that word-length is a continuous quantity and that the three-letter word category includes all words between 2.5 and 3.5 letters long, we have to add the 0.38 to the upper limit of the 3-letter interval.)

The procedure which we have just gone through to find the median for this particular distribution can be generalized and summarized in a formula:

$$\text{median} = L + \left(\frac{N/2 - cf}{f}\right)$$

where L is the upper limit of the greatest value whose cumulative frequency is less than 50% of the whole, cf is the cumulative frequency of that value, f is the frequency of the value next above it (the one containing the median in its range) and N is the total number of items in the distribution.

Exercise 5.

Calculate the median word-length of the passages from Hume & Macaulay in Exercise 1(a).

Solutions (i) $3\frac{19}{30} = 3.633$; (ii) $4\frac{7}{26} = 4.269$.

Given a frequency distribution, the mode is the easiest average to discover; but conversely, being told the mode gives one little information about the nature of the distribution. Mean and median each require calculation to discover, but each is more informative than the mode about the distribution in question. The median is a position or rank statistic: that is to say, it gives information about the ordering of the values in the distribution while being unaffected by their absolute values. The mean is affected by the size of each item in the distribution and is therefore, as observed, sensitive to extreme values. The median has the advantage that it can be calculated even where some of the classes used in the distribution are open-ended; if, for instance, in drawing up the distribution of the word-lengths in Table 2.1 the top three classes had been amalgamated into a single class ('words over 10 letters long') it would still be possible to calculate the median word-length but not the mean word-length. On the other hand, the mean can be used in further calculations, whereas the median often cannot. For example, given the median of each of the three 70-word passages we cannot compute the median of the whole 210-word passage of Mill. But given the mean of the segments we can calculate the mean of the whole. Given that the mean for the first 70-word section is 4.41, we can calculate that the total number of letters is 309; given that the mean for the second is 4.97 we calculate the total letters as 348, and given the mean for the third as 4.84 we reach the total of 334. This gives us a grand total of 996 for the 210-word passage, and dividing 996 by 210 we obtain the grand mean of 4.74. In general, the mean for a group can be found by multiplying the mean of each subgroup by the number of items in the subgroup, adding together these products, and then dividing by the sum of the number of items in each subgroup. This can be expressed in the following formula:

$$\bar{X} = \frac{N_1 \bar{X}_1 + N_2 \bar{X}_2 + \dots N_n \bar{X}_n}{N_1 + N_2 + \dots N_n}$$

which we may write in an abbreviated form as

$$\frac{\Sigma(N_i \bar{X}_i)}{N} \qquad (i = 1 \rightarrow n).$$

Where, as in the case of the Mill samples, the number of items in each group is the same, the calculation can be simplified: for the mean of the total group is equal to the mean of the subgroup means. It is easy to verify that (4.41 + 4.97 + 4.84)/3 = 4.74. The more complicated procedure described above, which is necessary when the number of items is not the same in each subgroup, is called *weighting* each average in accordance with the size of the group.

In a symmetrical distribution, the median and the mean will co-incide with each other (and with the mode, if the distribution is unimodal). In a positively skewed distribution the mean will be greater than the median; in a negatively skewed distribution it will be less. If we know the mode and the median of a distribution we can tell whether, and how, it is skewed.

Exercise 6.

Construct a frequency distribution and a cumulative frequency distribution of the length of sentences in words in the following passage. Calculate the mode, median, and mean sentence length. (Count as a single sentence each section of text beginning with a capital letter and ending with a full stop or question mark.)

Who can find a virtuous woman? Her price is far above rubies. The heart of her husband doth safely trust in her, so that he shall have no need of spoil. She will do him good and not evil all the days of her life. She seeketh wool and flax, and worketh willingly with her hands. She is like the merchants' ships; she bringeth her food from afar. She riseth also while it is yet night, and giveth meat to her household, and a portion to her maidens. She considereth a field, and buyeth it; with the fruit of her hands she planteth a vineyard. She girdeth her loins with strength, and strengtheneth her arms. She perceiveth that her merchandise is good: her candle goeth not out by night. She layeth her hands to the spindle, and her hands hold the distaff. She stretcheth out her hand to the poor; yea, she reacheth forth her hands to the needy. She is not afraid of the snow for her household: for all her household are clothed with scarlet. She maketh herself coverings of tapestry; her clothing is silk and purple. Her husband is known in the gates, when he sitteth among the elders of the land. She maketh fine linen, and selleth it; and delivereth girdles unto the merchant. Strength and honour are her clothing; and she shall rejoice in time to come. She openeth her mouth with wisdom; and in her tongue is the law of kindness. She looketh well to the ways of her household, and eateth not the bread of idleness. Her children arise up, and call her blessed; her husband also, and he praiseth her. Many daughters have done virtuously; but thou excellest them all. Favour is deceitful, and beauty is vain; but a woman that feareth the Lord, she shall be praised.

Solution to Exercise 6.

Sentence-length in words	Number such f	Cumulative frequency cf	fx
20	1	22	20
19	1	21	19
18	2	20	36
17	2	18	34
16	2	16	32
15	2	14	30
14	3	12	42
13	2	9	26
12	2	7	24
11	1	5	11
10	2	4	20
9	0	2	0
8	0	2	0
7	0	2	0
6	2	2	12
5	0	0	0
4	0	0	0
3	0	0	0
2	0	0	0
1	0	0	0
	$N = 22$		$fx = 306$

Mode $= 14$; mean $= 13\frac{10}{11} = 13.91$; median $= 14\frac{1}{6} = 14.167$.

In the passage studied in Exercise 6, the longest sentence was of 20 words, and the shortest 6 words; in the passages studied earlier the word lengths have been between 1 and 17 letters. The difference between the highest and lowest value in a distribution is known as the *range* of a distribution. Where the range is greater than 20 it is better to group the data by combining adjacent values into classes instead of counting each separate value as a distinct class. When this is done each class interval will contain several values: for instance, sentences between one and five words long may constitute one class, those between six and ten another class, and so on. Each item in the distribution is assigned to the appropriate class; e.g. seven-word sentences and eight-word sentences alike are assigned to the 6–10 class interval. Obviously, class intervals have to be defined so that they are mutually exclusive and jointly exhaustive of the possible values taken by the data.

If we group data in this way, the sentence length distribution for Exercise 6 would look as follows:

Sentence-length in words	Frequency f	Cumulative frequency	fx
16–20	8	22	144
11–15	10	14	130
6–10	4	4	32
1–5	0	0	0

How is the fx column calculated? What is the length of a sentence in the 16–20-word category? The answer is that each sentence in the category is assumed to have the value of the mid-point of the class: all sentences in the 16–20 class are assumed to be 18 words long, and all words in the 11–15 word class are assumed to be 13 words long.

It will be seen that the grouping of data results in a more compact distribution at the cost of a certain loss of information. Statistics calculated from grouped data will not be quite as accurate as those from ungrouped data. However, the grouping of data is practically inevitable for many purposes when the range of values is large. Most passages of consecutive prose, for instance, contain sentences very much longer than 20 words; hence in studying sentence-length distributions the data are commonly grouped. The class intervals in a grouping should be of uniform size, and the size of the interval should be such as to allow the total number of intervals to be less than 20. The size is determined by dividing the range by the number of intervals desired. The range of a set of sentences can be roughly estimated by looking out for what appears as the longest sentence in the text to be analysed and counting its length in words.

Consider, for instance, the following passage which commences Gibbon's *Autobiography*.

In the fifty-second year of my age, after the completion of an arduous and successful work, I now propose to employ some moments of my leisure in reviewing the simple transactions of a private and literary life. Truth, naked, unblushing truth, the first virtue of more serious history, must be the sole recommendation of this personal narrative. The style shall be simple and familiar: but style is the image of character; and the habits of correct writing may produce, without labour or design, the appearance of art and study. My own amusement is my motive, and will be my reward: and if these sheets are communicated to some discreet and indulgent friends, they will be secreted from the public eye till the author shall be removed beyond the reach of criticism or ridicule.

A lively desire of knowing and of recording our ancestors so generally prevails, that it must depend on the influence of some common principle in the minds of men. We seem to have lived in the persons of our forefathers; it is the labour and reward of vanity to extend the term of this ideal longevity. Our imagination is always active to enlarge the narrow circle in which Nature has confined us. Fifty or a hundred years may be alloted to an individual; but we step forward beyond death with such hopes as religion and philosophy will suggest; and we fill up the silent vacancy that precedes our birth, by associating ourselves to the authors of our existence. Our calmer judgement will rather tend to moderate, than to suppress, the pride of an ancient and worthy race. The satirist may laugh, the philosopher may preach, but Reason herself will respect the prejudices and habits which have been consecrated by the experience of mankind. Few there are who can sincerely despise in others an advantage of which they are secretly ambitious to partake. The knowledge of our own family from a remote period will always be esteemed as an abstract pre-eminence, since it can never be promiscuously enjoyed; but the longest series of peasants and mechanics would not afford much gratification to the pride of their descendant. We wish to discover our ancestors, but we wish to discover them possessed of ample fortunes, adorned with honourable titles, and holding an eminent rank in the class of hereditary nobles, which has been maintained for the wisest and most beneficial purposes, in almost every climate of the globe, and in almost every modification of political society.

Wherever the distinction of birth is allowed to form a superior order in the state, education and example should always, and will often, produce among them a dignity of sentiment and propriety of conduct, which is guarded from dishonour by their own and the public esteem. If we read of some illustrious line, so ancient that it has no beginning, so worthy that it ought to have no end, we sympathize in its various fortunes; nor can we blame the generous enthusiasm, or even the harmless vanity, of those who are allied to the honours of its name. For my own part, could I draw my pedigree from a general, a statesman, or a celebrated author, I should study their lives with the diligence of filial love. In the investigation of past events our curiosity is stimulated by the immediate or indirect reference to ourselves; but in the estimate of honour we should learn to value the gifts of Nature above those of Fortune; to esteem in our ancestors the qualities that best promote the interests of society; and to pronounce the descendant of a king less truly noble than the offspring of a man of genius, whose writings will instruct or delight the latest posterity. The family of Confucius is, in my opinion, the most illustrious in the world. After a painful ascent of eight or ten centuries, our barons and princes of Europe are lost in the darkness of the middle ages; but, in the vast equality of the empire of China, the posterity of Confucius have maintained, above two thousand two hundred years, their peaceful honours and perpetual succession. The chief of the family is still revered, by the sovereign and the people, as the lively image of the wisest of mankind. The nobility of the Spencers has been illustrated and enriched by the trophies of Marlborough; but I exhort them to consider the *Fairy Queen* as the most precious jewel of their coronet. Our immortal Fielding was of the younger branch of the Earls of Denbigh, who drew their origin from the Counts of Hapsburg, the lineal descendants of Eltrico, in the seventh century Duke of Alsace. Far different have been the fortunes of the English and German divisions of the family of Hapsburg: the former, the knights and sheriffs of Leicestershire, have slowly risen to the dignity of a peerage; the latter, the Emperors of

Germany and Kings of Spain, have threatened the liberty of the old, and invaded the treasures of the new world. The successors of Charles the Fifth may disdain their brethren of England; but the romance of Tom Jones, that exquisite picture of human manners, will outlive the palace of the Escurial, and the imperial eagle of the house of Austria.

Even on a quick perusal of this passage one sentence (that beginning 'In the investigation of past events...') stands out as exceptionally long. It is found to contain 79 words. Assuming, therefore, that no sentence in the passage contains more than 80 words we can choose four words as the interval for our classes, grouping sentences in classes of 1–4, 5–8, 9–12 and so on. In general we find that class intervals of four or five words suffice for all except the most prolix writers.

The sentence length distribution for this passage can then be set out as follows:

Sentence-length in words	Frequency f
77–80	1
73–76	0
69–72	0
65–68	0
61–64	0
57–60	2
53–56	0
49–52	2
45–48	2
41–44	2
37–40	2
33–36	1
29–32	4
25–28	2
21–24	1
17–20	3
13–16	2
9–12	0
5–8	0
1–4	0
	24

The grouping of data in this manner makes the frequency distribution clearer and facilitates the construction of histograms and frequency polygons. It does, however, make more complicated the calculation of the statistics of the distribution. We have to make the assumption that the cases falling within an interval are evenly distributed between its limits. This assumption will naturally, in most cases, be only an approximation to the truth.

On this assumption we can calculate the mean for the distribution on the basis that the mean sentence-length for all the sentences in each group is the mid-point of the group interval. Thus we assume that the mean length of the four sentences in the interval 29–32 is the mid-point of the group. Pretending, as we did in the case of word-length, that sentence-length is a continuous rather than a discrete quantity, we say that the real limits of the interval are 28.5 and 32.5, though naturally the only values which can actually occur are 29, 30, 31 and 32. The mid-point between the interval limits 28.5 and 32.5 is 30.5. We calculate in a similar manner for the other class intervals.

When we calculate the mean for the grouped data we multiply the number of cases occurring in each class by the mid-point of the group and then sum the products before dividing by the total number of items in the distribution. Thus we calculate the mean of the above distribution as follows:

Sentence-length in words	Mid-point of interval	Frequency f	Frequency × mid-point
77–80	78.5	1	78.5
73–76	74.5	0	0
69–72	70.5	0	0
65–68	66.5	0	0
61–64	62.5	0	0
57–60	58.5	2	117
53–56	54.5	0	0
49–52	50.5	2	101
45–48	46.5	2	93
41–44	42.5	2	85
37–40	38.5	2	77
33–36	34.5	1	34.5
29–32	30.5	4	122
25–28	26.5	2	53
21–24	22.5	1	22.5
17–20	18.5	3	55.5
13–16	14.5	2	29
9–12	10.5	0	0
5–8	6.5	0	0
1–4	2.5	0	0
		$N = 24$	$f \times mp = 868$

$$\text{Mean sentence length } \mu = \frac{(f \times mp)}{N} = \frac{868}{24} = 36\tfrac{1}{6} = 36.17.$$

The median is calculated as for ungrouped data, except in one respect. When we calculate the sum to be added to the upper limit of the interval below that containing the median, we have to take account of the fact that the interval size is no longer one. That is to say, instead of using the formula implicit in our previous calculation

$$\text{median} = L + \frac{(N/2 - cf)}{f}$$

we need to use the formula

$$\text{median} = L + \frac{(N/2 - cf)}{f} \times C$$

where C is the size of the class interval.

To compute the median of the passage from Gibbon we must first, as before, construct a cumulative frequency distribution:

Sentence-length in words	Cumulative frequency
77–80	24
73–76	23
69–72	23
65–68	23
61–64	23
57–60	23
53–56	21
49–52	21
45–48	19
41–44	17
37–40	15
33–36	13
29–32	12
25–28	8
21–24	6
17–20	5
13–16	2
9–12	0
5–8	0
1–4	0

In this case it can be seen immediately that the median must coincide with the upper limit of the 29–32 class, since 12 sentences, or one-half of the total, fall below that limit and 12 fall above it. The median is therefore 32.5. The same result is reached by mechanically applying the formula for the median, with the values 32.5 for L, 24 for N, 12 for cf, 1 for f, and 4 for C. The median will be

$$32.5 + \frac{(24/2 - 12)}{1} \times 4 = 32.5 + 0.$$

There is no simple and accurate way of calculating the mode for grouped data. Indeed, the calculation of mean and median for grouped data is a tedious and approximative matter. There are various devices for abbreviating the process which were taught in older textbooks of statistics. Recent textbooks tend to omit them since, with the advent of cheap calculators, it is no longer necessary to perform long calculations by hand and the original ungrouped data can be used with an increase of accuracy. The recording of such features of texts as sentence-length is likewise much more easily done by computer on machine-readable texts. Consequently a knowledge of the most expeditious methods of performing long calculations by hand upon masses of data is no longer necessary for the student.

Exercise 7.

Find the mean and median sentence-length of the 50-sentence passage whose sentence-length distribution is given below. (Gibbon, *Decline and Fall of the Roman Empire*, the passage from Chapter XVI beginning 'It was in the choice of Cyprian...', on p. 103 of volume 2 of Bury's edition.)

Sentence-length in words	Number such (f)
81–85	1
76–80	0
71–75	2
66–70	1
61–65	1
56–60	4
51–55	5
46–50	3
41–45	7
36–40	3
31–35	5
26–30	9
21–25	4
16–20	3
11–15	1
6–10	1
1–5	0

Note that the interval length is now five, not four as in the example worked in the text. Since the interval length is odd, the mid-points of the interval will occur at whole-number points instead of between them.

Solution to exercise 7.

Mean sentence length = 40 words; median sentence length = $38\frac{5}{6} = 38.83$.

4

Measures of Variability

I<small>F</small> <small>WE</small> know the mode, the median, or the mean of a distribution we are able to characterize the distribution by locating its centre. But in addition to these measures of central tendency we need to know, for even the most summary description of a distribution, something about the spread or dispersion of the items in the distribution around the central values. Two authors, for instance, may write sentences of approximately the same mean length and yet may have very different habits in respect of sentence-length. Perhaps one of them writes sentences of uniform length close to the mean value, while the other alternates very long sentences with very short ones. The effect of the two styles on the reader would be very different; and if we drew frequency polygons for the two distributions they too would look quite different. The polygon for the first author would be shaped like this

and the polygon for the second author would be shaped like this

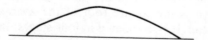

In order to make a compact numerical representation of this kind of difference we need a quantitative measure of the scatter or dispersion of values in the distribution. Such a measure is called by statisticians a measure of variability. Any distribution can be concisely described

51

by assigning it a measure of central tendency and a measure of variability.

Just as there are three measures of central tendency, the mode, the mean and the median, there are three commonly used measures of variability, the range, the standard deviation and the interquartile range. We will deal with each of these in turn.

The range is the easiest measure of variability to calculate, just as the mode is the easiest measure of central tendency to ascertain. In discussing class intervals for grouped data earlier, we said that the range of a set of values was the difference between the highest and lowest values. Strictly, it is the difference between the upper real limit of the interval in which the highest value occurs and the lower real limit of the interval in which the lowest value occurs: to take account of this, one unit of measurement must be added to the difference between the largest and smallest value. Thus the range of the word-length distribution for Mill on p. 36 above is 12; the range of the sentence-length distribution for the biblical passage in Exercise 6 is 15.

Just as the mode is the crudest and least informative measure of central tendency, so the range is the crudest and least informative measure of dispersion. It gives no information about the spread of the values over the range, and its value is determined entirely by the maximum and minimum values, so that if they are exceptional in some way it will give a misleading picture of the distribution as a whole. A more accurate picture will be given by a measure of dispersion which is based on every single item in the distribution.

Such a measure is the *standard deviation*, which is the measure of variability which corresponds to the mean as a measure of central tendency. It is indeed based on the mean, for it measures the dispersion of the items in terms of their deviation from the mean value. The standard deviation is the square root of the average squared deviation from the mean. It is calculated in four stages: first, the deviation of each item in the distribution from the mean is noted; secondly, each of these deviation scores is squared; the mean of these squared deviations is calculated; the square root of this mean is then calculated, and that is the standard deviation of the distribution. The definition of the standard deviation, and the stages of its calculation, are expressed in the following formula

$$\sigma = \sqrt{\frac{\Sigma(X - \mu)^2}{N}}$$

where

σ = standard deviation,
$\Sigma(X - \mu)^2$ = the sum of the squared differences between each item and the mean μ,
N = the number of items in the distribution.

If the deviation score is written x, an alternative way of writing the formula is:

$$\sigma = \sqrt{\frac{\Sigma x^2}{N}}.$$

Let us work through the calculation of a standard deviation in a simple instance. Consider the following ten lines from Keats' *Endymion*:

Leading the way, young damsels danced along,
Bearing the burden of a shepherd's song
Each having a white wicker over-brimm'd
With April's tender younglings: next, well trimm'd
A crowd of shepherds with as sunburnt looks
As may be read of in Arcadian books;
Such as sat listening around Apollo's pipe,
When the great deity, for earth too ripe,
Let his divinity o'erflowing die
In music, through the vales of Thessaly....

Let us suppose we wish to discover the mean and standard deviation of the length of the lines in words. As the whole passage contains 70 words (counting hyphenated words as single words) in 10 lines, it is easy to work out that the mean length is seven words. To work out the standard deviation we first record the deviation of each line's length from the mean length, thus:

Line	Length	Deviation
1	7	0
2	7	0
3	6	−1
4	7	0
5	8	+1

6	8	+1
7	7	0
8	8	+1
9	5	−2
10	7	0

These deviations will form the basis of the measure of dispersion: but we cannot simply add them together with a view to forming an average deviation. For the sum of the positive deviations from the mean exactly equals, and thus cancels out, the sum of the negative deviations. This will always be so and follows from the definition of the mean. For this reason we use not the deviation scores themselves, but their squares, as the basis of our measurement. The squares of the deviations will all be positive and will not cancel each other out. The squaring will also have the effect of giving extra weight to substantial deviations from the mean by comparison with minor divergences.

For the lines in our example the squared deviations are

$$0 \quad 0 \quad 1 \quad 0 \quad 1 \quad 1 \quad 0 \quad 1 \quad 4 \quad 0$$

making a total of eight. We divide this by the number of lines (10) in order to achieve the mean squared deviation of 0.8. It remains only to take the square root of this number to reach the standard deviation, which is $\sqrt{0.8}$ or 0.894.

The mean squared deviation, the penultimate stage in working out the standard deviation, is important enough to have a name of its own: it is called the *variance* of the distribution. Obviously, it contains exactly the same information as the standard deviation: given the standard deviation you can always work out the variance and vice versa. But the standard deviation is a more convenient statistic to work with for several reasons. One is that it is in the same units as the items in the distribution: the standard deviation of the line length in words is 0.894 *words*. But if we were asked to specify the unit of the variance, we would have to say 'words squared' and of course there is no sense to be made of such a unit. If we square an inch we get a square inch; but we cannot square a word and get a square word.

The method by which we calculated the standard deviation for the passage from Keats was by following the formula which corresponds to the definition of the standard deviation as the root mean squared deviation. There is another formula which is algebraically equivalent, which avoids the necessity for calculating the mean and the individual deviations. The formula is:

$$\sigma = \sqrt{\dfrac{\Sigma X^2 - \dfrac{(\Sigma X)^2}{N}}{N}}.$$

We can apply this formula to the data from the Keats passage.

X	X^2
7	49
7	49
6	36
7	49
8	64
8	64
7	49
8	64
5	25
7	49
$\Sigma X = 70$	$\Sigma X^2 = 498$

$$= \sqrt{\dfrac{498 - \dfrac{4900}{10}}{10}} = \sqrt{\dfrac{498-490}{10}} = \sqrt{0.8} = 0.894.$$

When the mean is a whole number, this is not as easy as the previous methods; but this is because the example is artificially simple. In most cases, the second method is simpler.

Exercise 8.

(a) Find the mean and standard deviation of the word-length in letters of the following passage:

Is this problem rather difficult? Never worry: integers assist solution.

(b) Find the mean, variance and standard deviation of the sentence-length in words of the following passage:

Fools! There is a time for rejoicing and there is a time for mourning and weeping. Some men live. Some men die. No man lives for ever. Who knows which of us will be alive tomorrow? The sun shines on the living and shines on the dead. Let us live while the sun shines. While the sun shines let us die. Better die young in glory than die old unwept.

Solution to Exercise 8.

(a) Mean six letters; standard deviation two letters.

(b) Mean seven words; standard deviation four words; variance 16.

In the examples above the mean and standard deviation were worked out without the values of the individual items being previously arranged into a distribution: we simply listed the length of each line in the order in which it came in the passage quoted. If the data are presented in the form of a frequency distribution, then the formula is as follows:

$$\sigma = \sqrt{\frac{\Sigma f X^2 - \dfrac{(\Sigma f X)^2}{N}}{N}}$$

where f represents the frequency of each value. Thus the sentence length distribution for the passage in Exercise (8b) can be laid out thus:

Sentence-length in words (X)	Number such (f)	$f X$	$f X^2$
15	1	15	225
11	1	11	121
9	2	18	162
7	2	14	98
5	1	5	25
3	2	6	18
1	1	1	1
	10	70	650

Since N has the value 10, and $\Sigma f X^2 = 650$ and $(\Sigma f X)^2 = 4900$ the value is

$$\sigma = \sqrt{\frac{650 - \dfrac{4900}{10}}{10}} = \sqrt{\frac{160}{10}} = 4.$$

When the frequency distribution contains groupings, the formula is the same except that the raw score X is replaced by the mid-point of the interval. The table above can indeed be taken as an illustration of this, with nine being regarded as the mid-point of the interval whose real limits are 8.5 and 9.5 and so on.

The value of a standard deviation, as was explained before, can never be negative. Its minimum value is zero: in a distribution with a zero standard deviation, every value in the distribution would be equal to the mean and to every other value. Thus a passage consisting of regular iambic pentameters would consist of lines whose mean length was five feet with a standard deviation of zero feet.

The standard deviation is measured in the same units as the items in the distribution. But besides providing a measure of dispersion for the entire distribution, the standard deviation can be used to measure the location of each individual item with respect to the whole. The value or score of an individual can be located in terms of its distance above or below the mean in comparison with the other individuals in the distribution. This is done by stating how many standard deviations above or below the mean the value is located. This is called the individual's z-value or z-score.

The z-score is the deviation of the raw score or value from the mean, divided by the standard deviation. Thus the formula is

$$z = \frac{X - \mu}{\sigma}$$

where, as usual, μ is the mean of the distribution and σ its standard deviation. If the raw value is above the mean, the value of z will be positive; if below, it will be negative. Thus, the individual is located by its distance in standard deviations above or below the mean: the mean is the reference point and the standard deviation the unit of measurement. z-Scores are not, like standard deviations, measured in the same units (inches, pounds, letters, words) as the items in the original distribution. A z-score, being the ratio of two quantities with (usually) physical dimensions is itself dimensionless: it is measured not in units of physical measurement but in the form of standard deviation units. z-Scores are therefore applicable to distributions of many different kinds. Consider two examples from the previous exercise. In (a) the word 'integers' contains eight letters, which is two more than the mean; since the standard deviation of the distribution is two letters, the word 'integers' has a z-score of $+1$. In (b) the mean is seven words, and the standard deviation four; the sentence 'The sun shines on the living and shines on the dead', which has 11 words, therefore likewise has a z-score of $+1$.

If we take all the values in a distribution and convert them into z-scores, we will have a new distribution of exactly the same shape as the original distribution, but with a mean of 0 and a standard deviation of 1. This is a simple consequence of the way in which z-scores are calculated. Distributions which have been standardized in this way are easier to compare with each other than distributions with values recorded in the original units. Thus the standard deviation not only provides a measure of scatter or variability, it also provides a common unit of measurement for the comparison of distributions of the most various kinds.

The third measure of dispersion which is commonly used, in addition to the range and standard deviation, is the interquartile range. This has the same role among the measures of variability as the median has among the measures of central tendency, and it is calculated in a very similar manner. The median was the point in the distribution which divided the values into two equal groups: the *quartiles* are the points which divide the values into four equal groups. The first quartile is the point below which 25% of the items in a distribution fall; the third quartile is the point below which 75% of the distribution items fall. (The first quartile, thus, is sometimes also called the 25th percentile, and the third quartile the 75th percentile.) The median is itself the second quartile and the 50th percentile. Median, quartiles and percentiles are all calculated on the basis of the cumulative frequency distribution. The distance between the first and third quartile is called the *interquartile range*. When the items in a distribution are well spread out the interquartile range will be great; if the values cluster closely around the median, the interquartile range will be less. The interquartile range gives more information about the spread of variables within a distribution than does the simple range. It is based on less information than the standard deviation and is, therefore, in general not as useful a statistic; however, it is not affected, as the standard deviation is, by the occurrence of a few extreme maverick values. In the case of distributions that are very skew it may be a less misleading statistic than the standard deviation; in general, when the median is the most appropriate measure of central tendency the interquartile range may be the most appropriate measure of dispersion.

We may illustrate the calculation of the interquartile range, as we illustrated the calculation of the median, from the sentence-length distribution of the passage from Gibbon's autobiography on p. 43–7 above. It will be recalled that the formula we adopted for the median was

$$\text{median} = L + \frac{(N/2 - cf)}{f} \times C.$$

The formula for calculating the quartiles is the same, except that $N/2$ is replaced by $N/4$ for the first quartile, and $3N/4$ for the third quartile. There is no need to use the formula to calculate the first quartile in this case, since it can be seen with the naked eye that six of the values, or one-quarter of the whole ($N = 24$) fall below the upper limit of the 21–24 class, i.e. 24.5, which is therefore the first quartile. But the point below which 18, or $\frac{3}{4}$ of the cases lie, it not so easily seen: the cumulative frequency merely tells us that it lies within the interval 45–48 (since 17 cases are below the 44.5 limit and 19 cases are below the 48.5 limit). If we apply the formula we obtain:

$$\text{third quartile} = 44.5 + \frac{(18 - 17)}{2} \times 4 = 44.5 + 2 = 46.5$$

where the value of L, the upper limit of the next highest interval below, is 44.5, that of N the items in the distribution is 24, that of cf (the cumulative frequency of the interval below) is 17, that of f (the number of items in the interval 45–48 itself) is 2, and that of C (the class interval size) is 4.

Since the first quartile is 24.5 and the third quartile is 46.5 the interquartile range is 22. Like the standard deviation, the interquartile range is in the same units as the distribution scores themselves: in this case it is 22 letters. The interquartile range can be divided by two to give the semi-interquartile range, which is called the quartile deviation and is sometimes used instead of the interquartile range as a measure of dispersion. In the present example the quartile deviation is 11 letters.

By specifying the first and third quartile and the median of a distribution we can give not only a measure of the dispersion of the items, but also an indication of the degree and nature of skewness. If the distribution is symmetrical, the two quartiles will be equidistant from the median; if it is skew, they will differ from it by different amounts. If the distribution is positively skewed, the distance between the first quartile and the median will be less than that between the median and the third quartile; if it is negatively skewed it will be the distance between the first quartile and the median which will be greater. If a distribution is extremely skew, it may be valuable to calculate, as an indicator of the extreme values, the 10th or 90th percentiles as well as the quartiles. They are calculated in a precisely parallel manner. (Just as the 25th percentile is called a quartile, so the 90th percentile may be called the 10th decile.) Because sentence-length distributions are commonly markedly positively skewed, they are frequently presented by specifying the mean, the median, the first and third quartiles and the ninth decile.

Exercise 9.

Calculate the third quartile and the ninth decile for the sentence-length distribution of the passage from Gibbon's *Decline and Fall* on p. 48 above.

Solution to Exercise 9.

Third quartile = 52.

Ninth decile = 60.5.

5

The Measurability of Literary Phenomena

FEATURES of style such as the length of words and sentences which an author uses, or the number of words in a poet's lines, are quantifiable in a straightforward and unproblematic way. There are other features, like choice of vocabulary or preferences between different parts of speech, which do not offer themselves for measurement in the same immediate and intuitive way, and it takes some reflection to see how numerical values can be assigned to these features as well. The problem is not peculiar to the statistical study of literature: it has been faced and met by statisticians working in biology and psychology who have long developed techniques for applying quantitative methods to the study of distinctions which are primarily qualitative.

Statisticians commonly distinguish between four different types of scales of measurement: ratio scales, interval scales, ordinal scales and nominal scales. We may consider these scales in turn and see in what way they can be applied to literary matters.

The *ratio scale* is the paradigm of a scale of measurement, and the measurement of length is the most obvious instance of a ratio scale. We may measure length in feet or metres or any other unit of length, but whatever unit we choose the measurement will have a number of features which are so important that they underlie all our procedures with measurements of length, and yet so obvious that they normally pass unremarked. First of all, the zero point on any scale of measurement is the same: no inches is the same length as no centimetres or no miles. Secondly, once we have chosen a unit each unit on the scale is the same as each other such unit: the distance between the 1-inch and the 2-inch mark on a rule is the same as the distance between the

61

3-inch mark and the 4-inch mark; each centimetre is just as long as every other centimetre. Consequently, something which is $2n$ units long is twice as long as something which is n units long; and this is true no matter what units the length is given in. It is this familiar fact which gives the ratio scale its name: no matter what is the unit of measurement, the ratio between the lengths of two given measured objects will be the same. If A is twice as long as B, measured in inches, it will also be twice as long as B in metric measurement.

An *interval scale* differs from a ratio scale in the following respect: its zero point is not determined by the nature of the measurement, but is fixed arbitrarily. Instances of interval scales are the Centigrade and Fahrenheit thermometer scales and the system of dating years according to the Christian era. The zero on the Centigrade thermometer is not the same point as the zero on the Fahrenheit thermometer, and neither of them is an absolute zero. Because the Centigrade zero is not absolute, we cannot say that a temperature of 100° Centigrade is twice as hot as a temperature of 50° Centigrade. Because the supposed date of Christ's nativity is not an absolute beginning of time we cannot say that the world will be twice as old in 2000 AD as it was in 1000 AD. If we were to translate these figures into the Fahrenheit scale or the Jewish calendar the appearance that the second figure of the pair represented twice as much of something as the first would disappear. On an interval scale the ratios between measurements in different units are not preserved.

However—and this is what gives the scale its name—the intervals between units of measurement are equal and equalities between measurements are preserved when the units of measurement are transformed. If the interval between A and B on the Centigrade scale is equal to the interval between B and C on the Centigrade scale, then the intervals between the corresponding points on the Fahrenheit scale will be equal. The length of time between 1780 AD and 1880 AD is the same as that between 1880 AD and 1980 AD no matter what calendar is used to reckon it. The distances between numbers on an ordinal scale are meaningful, though the ratios between them are not. Consequently, though multiplication cannot be applied to the numbers on an ordinal scale, addition and subtraction can. The large majority of statistical procedures—including all those included in the present work—are equally applicable to data measured on interval scales no less than to data measured on ratio scales.

The notion of an *ordinal scale* of measurement is an extension of the ordinary notion of measurement. When we record the order in which competitors pass the winning post, or rank the candidates for an office, or list the top ten gramophone records in order of popularity we can be thought of as assigning them to points on a scale. But if we simply record the order—rather than, say, the distance in time between the successive arrivals at the winning post—then we are assigning them to a scale on which only the order of the points, and not the intervals between them, have a significance. Hence the name 'ordinal scale'. Ordinal scales are frequently used in psychological and sociological studies; opinion polls are often invitations to rank politicians, proposals or products on an ordinal scale.

The crudest type of measurement, which is really not so much measurement as classification or codification, is measurement on a *nominal scale*. Wherever items can be assigned to different categories by virtue of some characteristic—persons as male or female, sentences as true or false, trees as oaks, ashes, beeches, etc.—we can arbitrarily assign numbers to the categories, putting, say, 1 for male and 2 for female, or 1 for true and 0 for false. The numbers here are not strictly functioning as numbers: they are really labels or names, which is why this 'measurement' is described as measurement on a *nominal* scale. Because they are not really numbers, the numbers occurring on a nominal scale cannot be added or multiplied or subjected to mathematical operations. But the objects or events assigned to the different categories are capable of being counted: the numbers counting the items in the different categories are perfectly genuine numbers. But the counting of objects in categories is quite a different operation from the measuring of objects in units and gives rise to different statistical procedures, as we shall see.

The use of numbers in nominal scales, though useful for coding purposes, is in a way misleading. For even the order of the numbers assigned has no significance. If oaks are assigned the number 1 and ashes the number 2, this implies no ranking of oaks vs ashes. On the other hand, ordinal ranking may take place without numbers being used. If several classes are grouped into lower, middle and upper, for instance, this is an assignment to an ordinal, and not just a nominal scale, in respect of social status.

How far does this typology for levels of measurement apply to the quantitative study of literary texts? We discover on investigation that every one of the levels of measurement, with its associated statistical procedures, can be applied to stylistic matters.

The study of the length of words in letters calls into play the characteristic features of a ratio scale of measurement. The limit of zero letters is an absolute, not a conventional, point of reference: we can thus say that a word of six letters is twice as long, in letters, as a word of three letters. However, there are two ways in which the measurement of the length of a word in letters differs from, say, the measurement of a person's height in centimetres. First of all, human height is a continuous variable, whereas the length of a word in letters is a discrete variable. A boy may be 95.5 centimetres tall, but a word can only be a whole number of letters long. Secondly, we cannot regard the length of a word as a single variable which can be measured by different units (e.g. words, syllables) as the height of a man is a single variable which can be measured in either inches or centimetres. If we measure the length of words in letters, then 'fright' is twice as long as 'pie'; if we measure in syllables the two words are of equal length.

Neither of these features, however, prevents us from treating such literary features as the length of a word in letters as being a quantity measurable on a ratio scale. Outside the literary field we have long been familiar with discrete variables such as family size: a family can have only a whole number of members yet we feel no qualms about saying that the average family size in the U.K. is 4.2. We have already seen in previous chapters the way which, in the construction of histograms and the calculation of elementary statistics, we can treat discrete variables as if they were continuous ones by treating the integer values of the variables as mid-points on a continuous scale. The length of words in letters, and the length of words in syllables, can each be treated as quantities measurable at the ratio level: but they have to be considered for what in fact they are—two independent variables and not a single variable measurable by two alternative units. They are related not as the height of a bookcase in inches and the height of a bookcase in centimetres, but as the height of a bookcase in inches and the height of a bookcase in shelves.

When we turn to consider the interval scale, it is not easy to find a literary characteristic appropriate for purely interval measurement. But we may very often wish to compare the variations in the stylistic features of an author's style with the progress of his life: and that progress is most naturally measured on the scale which provided the paradigm of an interval scale, the calendar of the era in which he lived.

Ordinal data are frequently encountered in a literary context. If we are told that Gibbon wrote longer sentences than Hume, and Hume longer sentences than Macaulay; if we are told that 'e' is the commonest letter in English, followed by 'o' and then by 't'; if we are told that in one set of texts nouns outnumber verbs, while in the other verbs outnumber nouns—in each of these cases we are being given points upon an ordinal scale. Ranking upon an ordinal scale may reflect an underlying distribution upon a ratio scale (as in the case of sentence-length) or a set of frequencies within categories of a nominal scale (as in the other two cases above); but a ranking may be possible, and permit conclusions, where no more precise data underlie the ranking. Thus a number of sensitive subjects may be asked to rank a series of passages of prose in order of their depressing quality. If there is a high degree of agreement between the independent rankings of different subjects, then the ordering may be said to have an objective value. But it could not be claimed to rest upon an underlying, more precise, quantitative determination of depressiveness.

The assignment of values on a nominal scale might be said to occur both in the most everyday and the most recherché activities of literary criticism. On the one hand, only somebody interested in coding a text to put into machine-readable form would find it worth while to replace the names of grammatical categories with numbers, putting say a '1' for a noun, a '2' for a verb, a '3' for an adjective and so on. On the other hand, every time we count the number of times a word occurs in a passage we are in effect performing the assignments into categories which are given effect by assignments to nominal values. If I count how many time 'thy' occurs in the Lord's Prayer, what I do is to assign each word in the prayer to one of two categories: either it is 'thy' or it is some other word. The number of 'thy's can be considered as the number of positive outcomes of a set of events, each of which has two possibilities.

To study an author's vocabulary by counting the number of times he makes use of particular words may seem a simple matter. However, there are a number of ambiguities in the notion of 'counting words' which need to be made explicit and resolved if any consistent significance is to be attached to word counts.

Words may be counted as *tokens* or counted as *types*. When a printer estimates the number of words that will fit on a page, or an editor commissions an article of 1000 words, they are counting words as tokens. When we compare the richness of vocabulary of two different writers, or ask how many words a schoolboy needs to know to understand a particular French unseen, we are counting words as types. The line 'Tomorrow, and tomorrow, and tomorrow' contains three tokens of the type 'tomorrow', and two tokens of the type 'and'. When we count occurrences of a word in a text, we are counting the number of tokens of a single type that occur. We are deciding, of each word-token in the text, whether or not it is an instance of the type in which we are interested. But the notion of 'type' is itself ambiguous. If our text contains the tokens 'do', 'does' and 'did', do we count these as tokens of three separate types or as three instances of a single word, the verb 'do'? We are free to take either course, so long as we make clear what we are doing. If we take the first course, we are making an *unlemmatized* word count; if we take the second course, we are *lemmatizing*, that is grouping forms which are typographically distinct under a single dictionary entry. Unlemmatized word counts are much easier to obtain by computer from machine-readable texts; they also contain more information, in the sense that it is always possible, with some labour, to construct a lemmatized word-count from an unlemmatized one, but an unlemmatized one cannot be recovered from a lemmatized one. But there are contexts in which a lemmatized one might be the more appropriate; and certainly a lemmatized word-count corresponds more closely to our intuitive idea of a quantitative study of vocabulary.

When we count words, then, we have to be clear what is meant by 'words'. We also have to consider what kind of counting we are doing. Raw counts, by themselves, are unlikely to be very valuable pieces of information. The Greek word for 'not', for instance, occurs 37 times in St Paul's Epistle to the Galatians. If we are to make any use at all of this information—say to make a comparison between the usage of this Epistle and other Epistles, or between St Paul's habits and those of other writers—we need at least to know how long the Epistle to the Galatians is, how many words it contains altogether. For most purposes what we are interested in is not the absolute frequency, or raw score, for a word, but its relative frequency or rate of occurrence. We want to know what *proportion* of the total tokens in a text are tokens of the type we are counting. In the case in point, since the total number of words in Galatians is 2220, the proportion of the text constituted by 'not' is $\frac{37}{2220}$ or 0.017. This figure, the rate or relative frequency of occurrence of the word for 'not', is a much more useful one than the absolute number of occurrences, and permits comparison with the rates for the same word in other texts.

Because even the commonest words, like the word for 'and' in various languages, occur with a very low relative frequency, less than 0.1, (i.e. less than one in ten), it is customary, in order to avoid constantly writing noughts after the decimal point, to express relative frequencies in terms of rate per thousand words, or per hundred words ('per cent'). Thus the frequency of the word for 'not' in Galatians would be expressed as 17 per thousand words or 1.7%.

Obviously, if we ascertain the relative frequency of every word which occurs in the text, the total of the relative frequencies will be 1 or 100%. It will be found, in any text in any language, that a large proportion of the total text consists of occurrences of a small number of relatively high-frequency words. A study of Aristotle's ethical treatises, for instance, showed that 13% of tokens in the entire text were instances of the definite article, and that another 25% of the text was accounted for by 36 common conjunctions and particles.

It may seem superfluous to point out that the rate of occurrence of a word is of more significance than the absolute number of times a word occurs in a text. But some students of literature, while admitting this in general, seem in practice to be reluctant to take account of it when the number of occurrences is one or zero. In particular a mysterious veneration is sometimes accorded to words which occur just once in a corpus (*hapax legomena*). In reality the rate of occurrence of a dull common word in a text may be a much more significant feature than a single appearance of a rare and striking word.

The rate of occurrence of a word in a text is the proportion of the text constituted by tokens of that word. This is not the only type of proportion which may be of interest in a text. Languages often contain pairs of synonyms, or near-synonyms, and it may be a characteristic feature of an author's style that whenever he has occasion to use a word with a particular meaning he chooses one rather than the other of two synonymous ways of expressing the meaning. Mosteller and Wallace, in their study of the contributions of Hamilton and Madison to the *Federalist* papers, discovered that there were striking differences between the authors' preferences between the pairs 'while–whilst' and 'on–upon'. Hamilton used 'while' nearly three times as often as he used 'whilst'; Madison used 'whilst' six times for every once he used 'while'. Hamilton used 'on' and 'upon' about equally; in Madison 'on' is 34 times as frequent as 'upon'. The relations between such rates can obviously be derived, as they were in the Mosteller and Wallace study, from the rates of occurrence in the text as a whole. But in order to study synonym-preference it is not necessary to count the words in the entire text. It is sufficient to count the number of occurrences of each of the synonyms. We can give numerical expression to an author's preference between synonyms A and B by calculating the proportion as follows:

$$\frac{\text{number of occurrences of A}}{\text{number of occurrences of A} + \text{number of occurrences of B}}.$$

Obviously, if the author prefers A to B, then this proportion will be greater than 0.5, otherwise it will be equal to or less than 0.5.

It is not only an author's choice between synonyms which provides a usefully quantifiable feature of style. The choice between other pairs of alternatives, which have a related though not identical function, may do so too: the choice, say, between 'all' and 'any' or 'more' and 'most', 'a' or 'an', 'in' or 'into'. Thus, for instance, A. Q. Morton has observed that in a set of passages from Jane Austen, the proportion of 'an'/'a' + 'an' is 0.143, whereas in a set of passages from her imitator who completed *Sanditon* the same proportion is 0.312. In theory there is no reason why one should not study proportionate pairs of words of quite unconnected meanings (e.g. the proportion of 'horse' to 'never'); but it is unlikely that such a study would reveal interesting characteristics of authors.

Besides being interested in an author's choice of words we may wish to investigate the contexts, large or small, within which he makes use of a particular word, or the other words with which he tends to collocate it. We can once again make use of proportions to do so. For instance, we may calculate what proportion, of all occurrences of the word 'the', occur as the first word of a sentence; or how often the word 'be', when it occurs, follows the word 'to'. The study of the preferred positions, and the favoured collocations, of words in a text can reveal, with a minimum of statistical apparatus, stylistic features which, however humble in themselves, may be highly characteristic of individual authors.

If we wish to compare proportions characteristic of one text with proportions characteristic of another we may make use of a simple statistic, invented by A. Ellegård, called the *distinctiveness ratio*. Suppose that we have an author A whose word usage we wish to compare with that of a corpus B, drawn from the writings of one or more authors. If we have the rates of occurrences for words in both A and B we can calculate the ratio

$$\frac{\text{rate of occurrence in A}}{\text{rate of occurrence in B}}.$$

If the word is more common in A than in B, this distinctiveness ratio (DR) will be greater than unity; if it is more greatly favoured by B, the DR will be fractional. Thus, for instance, the rate of occurrence of 'upon' in texts of Hamilton is 0.324%; in texts of Madison it is 0.023%. If we define the distinctiveness ratio of the word from the point of view of Hamilton, we find that it is $0.324\%/0.023\% = 14.09$. This is an extraordinarily high DR: most authors writing in the same language at the same time use most words at roughly the same rate, so that the DR of most words in a comparison of one author with others is usually near to unity. (We could, of course, have defined the DR from the point of view of Madison; it would then have been 0.07, an extraordinarily low DR.) We may call the words which have a DR greater than unity when defined from Hamilton's point of view Hamilton's *plus-words*; those which have a DR less than one are *minus-words*. For reasons which will become clear later, it is only DRs which differ substantially from unity which mark genuinely discriminating characteristics: for instance, *plus-words* with a DR greater than 1.5, or *minus-words* with a DR less than 0.67. But in order to characterize the peculiarities of an author's vocabulary account must be taken of DRs far from unity in both directions: minus-words may be just as indicative as plus-words.

Because a distinctiveness ratio is simply a ratio between two proportions, we can work out DRs not only for the rates of occurrences of particular words, but also for synonym choices, collocations and all the other proportions listed above. The distinctiveness ratio of the proportion an/a + an for Jane Austen in comparison with her imitator is 0.45. The linguistic habit is thus an indicative minus-feature of her style in comparison with her imitator.

Exercise 10.

For each of the following two passages:

(a) Calculate the rate of occurrence, in per cent, of the words 'the' and 'and' and 'that' and 'their'.

(b) Rank, in descending order of frequency, the prepositions 'in', 'of', 'by', 'with'.

(c) Calculate the proportion 'a'/'a' + 'the' and the proportion 'they'/'they' + 'their'.

(d) Work out what proportion of occurrences of 'the' are not followed immediately by a noun.

(e) Calculate the DR of 'the', 'and', 'of' and 'that' and decide which are plus and minus words, in comparison with the other passage.

(f) Calculate the DRs of the proportions worked out in (c).

Sir James Tyrell devised that they should be murdered in their beds. To the execution whereof, he appointed Miles Forest, one of the four that kept them, a fellow fleshed in murder beforetime. To him he joined one John Dighton, his own horsekeeper, a big, broad, square, strong knave. Then, all the others being removed from them, this Miles Forest and John Dighton, about midnight (the silly children lying in their beds) came into the chamber and suddenly lapped them up among the clothes, so bewrapped them and entangled them, keeping down by force the feather bed and pillows hard unto their mouths, that within a while, smothered and stifled, their breath failing, they gave up to God their innocent souls into the joys of heaven, leaving to their tormentors their bodies dead in the bed. Which after that the wretches perceived, first by the struggling with the pains of death, and after long lying still, to be thoroughly dead: they laid their bodies naked out upon the bed, and fetched Sir James to see them. Which, upon the sight of them, caused those murderers to bury them at the stair foot, meetly deep in the ground, under a great heap of stones. Then rode Sir James in great haste to King Richard, and, showed him all the manner of the murder, who gave him great thanks, and, as some say, there made him knight. But he allowed not, as I have heard, that burying in so vile a corner... (Sir Thomas More, *The History of King Richard the Third*, ed. Campbell 1931, 451.)

... In what manner they perished was kept a profound secret; the following is the most consistent and probable account, collected from the confession made by the murderers in the next reign. Soon after his departure from London, Richard had tampered in vain with Brakenbury, the governor of the Tower. From Warwick he despatched Sir James Tyrrel, his master of the horse, with orders that he should receive the keys and the command of the fortress during twenty-four hours. In the night Tyrrel, accompanied by Forest, a known assassin, and Dighton, one of his grooms, ascended the staircase leading to the chamber in which the two princes lay asleep. While Tyrrel watched without, Forest and Dighton entered the room, smothered their victims with the bed-clothes, called in their employer to view the dead bodies, and by his orders buried them at the foot of the staircase. In the morning Tyrrel restored the keys to Brakenbury, and rejoined the king before his coronation at York. Aware of the execration to which the knowledge of this black deed must expose him, Richard was anxious that it should not transpire; but when he understood that men had taken up arms to liberate the two princes, he suffered the intelligence of their death to be published, that he might disconcert the plans, and awaken the fears of his enemies. The intelligence was received with horror both by the friends and the foes of the usurper; but it did not dissolve the union of the conspirators. (Lingard, *History of England*, Vol. IV, p. 243.)

Solution to Exercise 10.

More	Lingard
(a) the: 5.6%	(a) the: 13.2%
and: 2.8%	and: 2.8%
that: 2.0%	that: 1.2%
their: 2.8%	their: 1.2%
(b) in/of/by/with	(b) of/in/with/by
(c) 2/9 = 0.222	(c) 2/35 = 0.057
3/7 = 0.429	1/4 = 0.25
(d) 2/7 = 0.286	(d) 4/33 = 0.12
(e) 14/35 = 0.42	(e) 33/14 = 2.36
1	1
3/8 = 0.375	8/3 = 2.67
5/3 = 1.67	3/5 = 0.6
(f) 35/9 = 3.889	(f) 9/35 = 0.257
12/7 = 1.714	7/12 = 0.583

6

Correlation and Bivariate Distributions

THE statistical procedures so far introduced have had as their purpose the description of a series of values of individual variables. The
distributions we have looked at have recorded the many values of a
single variable, such as word-length; the proportions we have studied
have been based on the many occurrences of a single word in a single
author. At the end of the previous chapter we introduced the distinctiveness ratio, which provides a means of comparing the occurrences
of a single word in more than one author. We have not yet encountered any method of comparing two different linguistic features in a
single author, or of putting the distribution of a set of words in one
author side by side with the distribution of the same set of words in
another author. In technical terms, we have confined our attention to
univariate distributions.

In the present chapter we turn to *bivariate* distributions: we shall
learn how to compare two different distributions and how to investigate relationships which may exist between two variables. We may
wish to explore, for instance, whether authors who use long sentences
also use long words. This would involve comparing sentence-length
distributions with word-length distributions. We might collect, for a
dozen or so authors, details of the mean sentence-length and the
mean word-length of their writings, and see whether those who
scored high on sentence length also scored high on word-length. This
is typical of the situation in which we study bivariate distributions:
we have available measurements on two variables for each of a set of
individuals and we try to find if there is any relationship between the
paired data. In the case suggested we would be comparing two
authors in respect of two different stylistic features; but we might also

73

compare the occurrence of words in two different authors. Here the words would be the individuals, and a word's frequency in author A would be one variable, and its frequency in author B the other variable. We would expect certain variables to go together somewhat: the length of word in letters would vary in rather the same way as the length of word in syllables for instance; on the other hand, the length of sentences would vary inversely with the number of full stops. In other cases we might not know: does age, for instance, vary with prolixity? In the present chapter we will set out statistical methods of studying such relationships.

As an illustration we can begin with a problem which concerns not exactly the relation between age and prolixity, but between age and the choice of length of words in poetic drama.

The following table reproduces data given in Charles Muller's *Etude de Statistique Lexicale: le Vocabulaire du Théâtre de Pierre Corneille* (1967). It gives, for 10 of Corneille's 32 dramas, the approximate date of composition and the mean number of words in each line of verse.

Drama	Date	Mean words per verse
Mélite	1629	8.93
La Galerie du Palais	1632	9.02
L'Illusion Comique	1635	9.15
Horace	1640	9.26
Rodogune	1644	9.15
Andromède	1650	9.20
Sertorius	1662	9.22
Agésilas	1666	9.32
Pulchérie	1672	9.48
Suréna	1674	9.53

This table can be regarded as locating each of the 10 individual dramas on two dimensions: that of time and that of word length (as measured by the number of words that can be fitted in to an alexandrine verse). We can represent the location of the plays in these two dimensions in a more graphic manner by constructing a diagram called a scattergram. This will provide a useful and visually satisfying way of indicating whether there is a correlation between the two variables, date of composition and number of words in the average verse.

The first step is to set out the two scales on the graph, the horizontal scale, or abscissa, to show date of composition; and the vertical scale, or ordinate, to show mean words per verse. (By convention, the values of the variable on the horizontal scale are called 'X values', and the values on the vertical scale are called 'Y values'.) To locate each play we find the point on the horizontal scale corresponding to its year of composition, and we then follow a vertical line from that point until we are level with the point on the vertical scale corresponding to its mean number of words. We then mark the spot we have reached with a dot or cross. The dot will mark the location of the play in both dimensions. Thus the first (leftmost) dot represents *Melite*, situated at 1629 on the time scale, and 8.93 on the word-length scale.

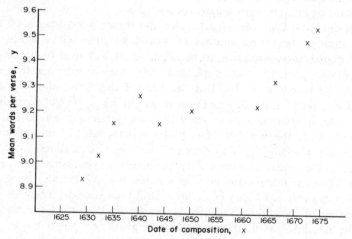

The graph shows that there is indeed a relationship between the date of composition of the plays and the average number of words per verse. In general, the further right a point is located along the horizontal scale, the higher up it is located on a vertical scale; which represents that in general the later a play was written the more words there are likely to be in one of its verses.

The relationship between the values of one variable and the values of another variable is known as their *correlation*. When, as in the present case, high scores on one dimension go with high scores on another dimension, the two variables are said to be positively correlated. If high scores on one scale go with low scores on the other, the correlation is described as negative. Our graph shows us that as Corneille grew older, he packed more words into each line. This must mean that as he grew older, he tended to use shorter words (at least, if we measure the length of words in syllables). If we had plotted along the vertical scale the mean length of word in syllables in a verse, not the mean number of words, the dots on our graph, instead of rising from the left-hand side towards the right-hand side, would have sloped down from left to right. Such a downward slope across a graph is characteristic of a negative correlation, and in our case would indicate that in the case of the dramas we have listed lateness correlates negatively with mean word-length.

Suppose now that we were to plot the mean word-length of the plays against the mean number of words per line. Here again we would expect the correlation to be negative: it was part of the argument of the previous paragraph that the more words per line the shorter the words must be. But the line of the dots on the graph would not only slope downwards, it would be a perfectly straight line—at least if Corneille's alexandrines are perfectly regular. For in an alexandrine the number of syllables is fixed at 12; consequently the mean word-length of any play can be exactly calculated once we are given the mean number of words per line: you simply have to divide 12 by the mean number of words per line. Where the values of one variable can be exactly calculated from the others in this way, the two variables are said to be perfectly correlated. A perfect correlation may be negative, as in the example just given; or positive—as, for instance, the correlation between the number of syllables in a play and the number of alexandrines in a play.

Perfect correlation is not commonly found except where two variables are connected with each other by definition. Certainly such correlations are not to be found between literary variables except where, as in the examples above, they can be demonstrated *a priori*. Correlations discovered by empirical research are always less than perfect, and statisticians have devised a number of measures of the strength or degree of such correlations. These measures are known as *coefficients of correlation* and they take values ranging from +1 for perfect positive correlation, through 0 for a total lack of correlation to −1 for perfect negative correlation. A coefficient of correlation shows both the direction and the magnitude of correlation. The + or − sign shows whether the correlation is positive or negative; the distance from 0 in either direction shows whether the correlation is low or high. Strength and direction are independent of each other: both a coefficient of +0.95 and of −0.95 mark a high degree of correlation, but the former indicates positive correlation and the latter negative correlation. The + and − signs of the correlation coefficient correspond to the slope of the cluster of points in the scattergram as indicating direction; the numerical value of the coefficient corresponds to the tightness of the cluster as indicating the strength of the correlation. Where there is little or no relationship or correlation between two variables the points will be dotted all over the scattergram and the correlation coefficient will be close to zero.

The simplest coefficient of correlation to calculate is a rank correlation coefficient which is an ordinal statistic based on the ranking of the values of the two variables. The most commonly used one is called, after its inventor, the Spearman Rank Correlation Coefficient, and it is usually symbolised by the Greek letter rho. Spearman's rho is calculated in accordance with the formula

$$\rho = 1 - \frac{6\Sigma d^2}{N(N^2 - 1)}$$

where N is the number of items or individuals in the array being studied, d, is the difference between each item's rank on the two scales, and ρ is the coefficient of rank correlation.

To illustrate the calculation of Spearman's ρ let us first rank the Corneille plays along the scales of date and words per verse. We must then note the differences between each play's rank on one scale and its rank on the other scale. Each difference is squared and the squares are added together, and the formula is then applied to the sum of the squared differences. The rankings, the differences and their squares are set out in the following table.

Drama	X Rank	Y Rank	Difference d	d^2
Mélite	1	1	0	0
La Galerie du Palais	2	2	0	0
L'Illusion Comique	3	3	0	0
Horace	4	7	−3	9
Rodogune	5	4	+1	1
Andromède	6	5	+1	1
Sertorius	7	6	+1	1
Agésilas	8	8	0	0
Pulchérie	9	9	0	0
Suréna	10	10	0	0

It is easy to calculate from this table that the sum of the squared differences, d^2 is 12. Since the number of dramas in the ranking is 10, we can obtain the value of Spearman's ρ by substituting into the formula:

$$\rho = 1 - \frac{6 \times 12}{10 \times 99} = 1 - \frac{72}{990} = 1 - 0.073 = 0.927.$$

The value of 0.927 is a high one, and shows that in this series of 10 plays there is a strong positive correlation between lateness in chronological order and number of words in a line.

In the table above L'Illusion Comique was assigned the rank 3, and Rodogune the rank 4. In the earlier table the mean number of words per verse is given as 9.15 for each drama. It was possible to assign different ranks in this case because in the original data from which the table was simplified the mean verse-length of L'Illusion Comique was given as 9.147 and that of Rodogune as 9.152; both were rounded off to 9.15. Suppose that this information had not been available, how would the tie have been dealt with in calculating the coefficient?

Statisticians offer various methods of dealing with ties in ranking. One is to order the tied items in some random way, e.g. by tossing a coin or placing them in alphabetical order. A more popular method is to assign the tied items the mean of the ranks which they must occupy between them. Thus if two items tie for the nth place, we give them the value $n + 0.5$; if three items tie for the rank n, to assign them all the rank $n + 1$. If there are a large number of ties, however, Spearman's rho becomes an inappropriate measure of correlation, and another coefficient of correlation must be used, such as Pearson's which will be introduced later in the chapter.

Exercise 11.

On the basis of the following data, calculate Spearman's coefficient for the correlation between the lateness of a play of Shakespeare and the amount of rhyme it contains.

Play	Date (according to E. K. Chambers' chronology)	Rhymed lines as proportion of all five-foot lines
Comedy of Errors	1592	19
Midsummer Night's Dream	1595	43
Much Ado about Nothing	1598	5
Measure for Measure	1603	4
Macbeth	1605	6

(The data are taken from Campbell & Quinn, *A Shakespeare Encyclopedia*, pp. 112, 932.)

Solution: $\rho = -0.60$.

Spearman's ρ is a comparatively crude coefficient of correlation. It takes into account only differences in ranking between values on the two scales which measure the variables to be correlated. It takes no account of the differences in magnitude between one value and another. In the data for the exercise above, for instance, it is clear that the difference between the proportion of rhymed lines in the *Comedy of Errors* and *A Midsummer Night's Dream* (19% and 43%) is enormously greater than the difference between the proportion in *Much Ado About Nothing* (5%) and that in Macbeth (6%). Yet these differences, in the calculation of Spearman's ρ, appear equalized: each of them is simply a difference of one place in a rank order, in which *Midsummer Night's Dream* has the first rank and *Comedy of Errors* the second, and in which *Macbeth* is ranked third and *Much Ado* fourth.

A more sensitive measure of correlation is given by Pearson's product–moment coefficient, which takes account not only of the ranking of a set of values, but of the absolute differences between the values themselves. To understand how Pearson's coefficient of correlation is calculated, we have to reflect upon the nature of a bivariate frequency distribution.

Suppose that we have the following data for the occurrence of 10 frequent words in two texts.

Word	Occurrences in Text A	Occurrences in Text B
The	15	9
And	11	8
Of	9	8
To	9	7
In	7	6
Then	7	6
That	5	5
By	3	5
A	3	4
Be	1	2

We may wish to know whether the frequency of words in one text is correlated with their frequency in the other text. We can consider each word as an individual about whom we have information along two scales: occurrence in A and occurrence in B. We can consider the values along these two scales as values of two variables whose relationship can be investigated.

The Pearson coefficient will give us a numerical measure of the correlation between these two variables. In order to calculate it, we may first set out the data in the above table in a bivariate frequency distribution thus:

```
O    10
c
c     9                                              x
u
r     8                              x    x
r
e     7                         x
n
c     6                    xx
e
s     5          x    x
i
n     4          x
B
      3
      2    x
      1
(Y)        1  2  3  4  5  6  7  8  9 10 11 12 13 14 15
           Occurrences in Text A (X)
```

A bivariate frequency distribution can be regarded as including an amalgam of two univariate frequency distributions. If we read along the horizontal axis the values of X, and ignore the vertical axis, we can count off numbers of crosses above each value reading as giving the frequency corresponding to that value, just as in a univariate distribution. Similarly, if we ignore the horizontal axis, we can read the vertical axis, with the number of xs parallel to each point on the scale, as giving a univariate distribution in a histogram which is lying on its side. The table as presented above combines these two distributions into a single chart.

The distributions along each axis can be converted, like any other distributions, into standardized distributions in which the raw scores are converted into z-scores, that is to say, in which the items are measured in terms of their standard deviations from the means of the distribution. Let us do this as a first step towards calculating the Pearson product-moment coefficient.

The distribution along the X axis can be represented thus:

Number of occurrences in A	Number of words occurring X times in A
(X)	(f)
15	1
11	1
9	2
7	2
5	1
3	2
1	1

It is easy to calculate the mean of this distribution as 7 and the standard deviation as 4. Similarly, we could calculate the mean of the Y distribution as 6 and the standard deviation as 2. We can now use the standard deviations to convert the X- and Y-scores into standardized z-scores: instead of a score of 7 on the X scale we put 0, instead of 11 we put $+1$, and instead of 3 we put -1. If we were to take the trouble to redraw the scales on the bivariate frequency distribution so that it was calibrated in z-scores instead of in the raw scores of occurrences, it would look like this:

```
S
t
a
n
d     +2                                                    x
a
r     +1                                       x   x
d                                              x
S      0                                    xx
c                                  x   x
o     -1                           x
r
e     -2              x
s
(Y)  -1.5    -1  -0.5   0   +0.5   +1   +1.5   +2
             Standard scores (X)
```

The Pearson correlation coefficient is a measure of correlation based on the pairs of standard scores. The formula which defines it is

$$r = \frac{\Sigma z_x z_y}{N}$$

where r is the correlation coefficient, N the number of paired values, and $z_x z_y$ the product of the z-score on the X-scale and the z-score on the Y scale. Let us calculate the coefficient on the basis of the data above.

Word	z_x	z_y	$z_x z_y$
The	+2	+1.5	3
And	+1	+1	1
Of	+0.5	+1	0.5
To	+0.5	0.5	0.25
In	0	0	0
Then	0	0	0
That	-0.5	-0.5	0.25
By	-1	-0.5	0.5
A	-1	-1	1
Be	-1.5	-2	3

Summing the products of the z-scores we obtain the value 9.5. To get the correlation coefficient r we simply divide by the number of pairs (10) and reach the result $r = 0.95$: a high degree of positive correlation.

The positive correlation is already manifest in the table of z-scores. It will be seen that values above the mean on one scale are also above the mean on the other, and that the actual values in each pair are fairly close together. Had the correlation been negative, negative scores on one scale would have been paired with positive scores on the other, making the products of the z-scores themselves negative. If there had been little or no correlation, the scores in each pair would not have been close together. If the two variables had been perfectly correlated, then the numerical values of the z-scores in each pair would have been exactly equal; but if the perfect correlation had been a negative one, the signs of the two scores would have differed.

In the original data for this example, we gave the absolute occurrences of the words, and not their relative frequencies or rates. For purposes of calculating correlation coefficients, it is immaterial which form the data are presented in. To give the rates instead of the absolute frequencies would simply have the effect of multiplying every value in each of the pair of distributions by a constant (1/total number of words in text). This would not have had any effect on the z-scores and therefore would not have altered the resulting value of the correlation coefficient.

Exercise 12.

On the basis of the following data calculate Pearson's coefficient of the correlation between frequency in Text A and frequency in Text B for the five words listed.

Word	Occurrences in A	Occurrences in B
He	40	42
She	52	30
It	34	24
This	16	18
That	28	6

Solution: $r = 0.6$.

Calculation of the Pearson product–moment coefficient by means of its definition formula can be a slow and laborious business, since it involves calculating the mean and standard deviation for each of the two distributions involved, and the conversion of all the values along each scale into z-scores. This will have become apparent to the reader on working through even the artificially simple examples in the text and in the exercises. Textbooks of statistics offer various formulae which enable the Pearson coefficient to be calculated from the raw data. One such is as follows:

$$r = \frac{\Sigma XY - \dfrac{(\Sigma X)(\Sigma Y)}{N}}{\sqrt{\left[\Sigma X^2 - \dfrac{(\Sigma X)^2}{N}\right]\left[\Sigma Y^2 - \dfrac{(\Sigma Y)^2}{N}\right]}}.$$

For the values given in the 10-word example above, ΣX is 70, ΣY is 60; ΣXY is 496, ΣX^2 is 650 and ΣY^2 is 400. Substituting in the formula we get

$$r = \frac{496 - \dfrac{(70)(60)}{10}}{\sqrt{\left(650 - \dfrac{4900}{10}\right)\left(400 - \dfrac{3600}{10}\right)}}$$

$$= \frac{496 - 420}{\sqrt{(160)(40)}} = \frac{76}{80} = 0.95.$$

X	Y	XY	X^2	Y^2
15	9	135	225	81
11	8	88	121	64
9	8	72	81	64
9	7	63	81	49
7	6	42	49	36
7	6	42	49	36
5	5	25	25	25
3	5	15	9	25
3	4	12	9	16
1	2	2	1	4
70	60	496	650	400

The reader may care to work out the Pearson coefficient for the data in the previous exercise by this alternative formula. In practice nowadays Pearson's r is rarely computed by hand: even quite modest pocket calculators commonly include a routine for its calculation. There is no need, therefore, for the reader to be alarmed by the laboriousness of the application of the formula.

For the product–moment coefficient to be a useful coefficient of correlation, for our purposes two conditions must be satisfied. First, the variables must be values on an interval or ratio scale. Secondly, the relationship between variables must be a linear one. Let us explain these conditions in turn.

Where the variables to be compared are not measurable on an interval or ratio scale, but can only be ordered according to ranks, Spearman's ρ is a more appropriate index of correlation than Pearson's r. The same is true where, even though the magnitudes involved are theoretically measurable on an interval scale, only information about rank ordering is available. In such circumstances, the Spearman ρ will be a good approximation to the Pearson r, provided that there are not many tied ranks. If there are ties, the Spearman coefficient will underestimate the correlation. If there are more than one or two ties, the Spearman coefficient should be calculated not according to its own definition, but according to the raw-score formula for the Pearson coefficient, using the ranks as raw scores.

Statisticians have devised a number of specialized correlation coefficients for cases in which the variables on the X and Y scales are of quite different kinds: for instance, where one variable is measurable on an interval scale and the other is simply a division into one of two classes (e.g. male vs female). These will not be considered in the present work but can be found in many statistics textbooks.

Where both variables occur on a nominal scale—that is to say, where the 'measurement' of the variable consists simply in classification into two or more classes—there are areas of statistics dealing with what are called attribute tests and contingency tables, which provide appropriate techniques which will be described later in the book.

Finally, and very importantly, the relationship between two sets of data can only be measured by the Pearson coefficient if the relationship is a linear one. A linear relationship is one which corresponds to a straight line, as opposed to a curved line, on a graph: a relationship in which similar relative increments on one scale are accompanied by similar relative increments on the other. Scattergrams where a curved or crooked line matches the data better than a straight one are said to represent *curvilinear* relationships, and more complicated methods

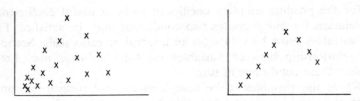

are needed to give a numerical value to the relationship. The diagrams show two types of scattergram where a Pearson coefficient would be an inappropriate measure of the relationship between two variables: in the left-hand one it would be inappropriate because the scatter of the values becomes wider and wider towards the right; in the right-hand one it would be inappropriate because the relationship is clearly not linear.

An obvious example of a non-linear relationship would be given by a graph which plotted along the X-axis miles travelled northward along the M6 and along the Y-axis distance from Manchester. At first as you travel northwards every mile travelled is a mile nearer to Manchester; but once you pass Manchester the distance grows with every further mile travelled. The graph would resemble the right-hand diagram above.

Such a relationship is a perfect non-linear relationship. Curvilinear relationships which are less than perfect would produce a scattergram which was crescent- or banana-shaped. In human beings, the relationship between age and various abilities is curvilinear. Strength, for instance, increases up to a certain point and later decreases after a period of stability. The same is true of a number of mental abilities associated with writing and creativity. Plotting certain features of literary style against the author's age might therefore often be expected to produce a curvilinear rather than a linear graph.

It will by now be clear that a scattergram is not simply a visual aid to make more vivid a correlation which is represented more precisely by the calculation of a correlation coefficient. It is often a necessary preliminary step to make clear whether the calculation of a linear correlation coefficient is an appropriate way to measure relationship.

It is a commonplace of elementary statistics that a correlation between two variables does not mean that there is a causal relationship between them. In order to establish that smoking was a cause of lung cancer—rather than, say, the effect of a genetic factor which also predisposed to cancer—it was not sufficient to observe the correlation between smoking habits and the incidence of cancer; it was necessary to produce cancer in the laboratory.

In the case of the variables to be measured in a literary text there is little temptation to see the relationship between one phenomenon and another in terms of a simple cause-and-effect relationship. If a tendency to use long words is found to go with a tendency to use long sentences, this is unlikely to be interpreted as a relationship between the two variables; the two characteristics are more likely to be seen as joint indices of a preference for a rotund style. If the frequency of words in one author correlates very highly with the frequency of those words in an other, this need not mean that one is imitating or echoing the other; it is likely to be due to the constraints of the language which they have in common.

The interpretation of a correlation is not a matter for the statistician as such. If an unexpectedly low correlation is found between two texts allegedly by the same author, it is for the literary critic or the historian, not for the statistician, to say whether this is an indication of diverse authorship or of a decision on the part of one and the same author to vary his style.

There is, however, a preliminary question which it is the statistician's business to answer. That is, whether *any explanation at all* is to be sought for a particular observed correlation. The reader may have wondered what, if any, significance was to be attached to some of the correlations worked out in the text or in the exercises. To what extent could similarities in frequencies of words, or relationships between other stylistic features, be regarded as simply a matter of chance coincidence?

A large part of the science of statistics is devoted to answering questions of this kind: to answering questions about the *statistical significance* of observed or calculated data. Statistical significance is not the same thing as significance *tout court*. A variation in word use between two different works of the same author may be statistically significant without being in the least bit significant in the history of his *oeuvre*. Statistical significance, then, is not a sufficient condition of literary or historical significance, nor is it a necessary condition in all cases: but if a phenomenon lacks statistical significance this may put the student on his guard against attaching too great importance to it. It is idle for the critic or the historian to seek causes for a variation in an author's habits when chance is an adequate explanation of the observed variations. It is the statistician who can indicate the bounds within which it is reasonable to seek explanations other than chance.

Whether a correlation coefficient, for instance, is significant depends on two things: on the number of pairs of items on which the correlation is based, and the numerical value of the coefficient (positive or negative). If we decide that we will regard a result as significant if the odds against its having been a chance coincidence are 20–1, then we can construct the following table of the significance of a coefficient of correlation. The table applies to either the Pearson or the Spearman coefficient.

Number of pairs in correlation	Lowest significant numerical value
3	0.99
4	0.95
5	0.88
6	0.81
7	0.76
8	0.71
9	0.67
10	0.63
15	0.51
20	0.44

This table will enable the reader to go back and evaluate which of the coefficients of correlation we have worked out were statistically significant, and it will provide a yardstick of significance for future correlations which he may work out himself. But in order to explain the principles on which such a table can be constructed, and what exactly is the information it contains, we must turn to the topic of inferential statistics, which is the concern of the second part of this book.

7

From Sample to Population

IN THE first part of the book we have been concerned with the methods by which statisticians summarize, analyse and compare sets of data. We have learnt how to compute statistics such as the mean, the standard deviation and the coefficient of correlation, with a view to presenting in a clear and concise manner the information actually contained in the data presented. The use of statistics in this way is known by statisticians as the practice of *descriptive statistics*.

Descriptive statistics is only a small part of the overall discipline of statistics. Most commonly, a statistician is concerned not simply to abbreviate or interpret information already present in actually observed data: he wishes also to make use of the information in his data to make well-founded conjectures about items which have not been directly studied but which resemble in crucial respects the observed phenomena. The data available are treated as a *sample* of a larger *population*, and inferences are drawn from the features of the sample to features of the population as a whole. The correct procedures for drawing inferences about the population as a whole from knowledge provided by a sample form the major part of the branch of statistics known as *inferential statistics*.

When a pollster is sampling opinion on a political issue, the population which he is aiming to study is quite literally a population of actual human beings in a specific area. But the word 'population' is used by statisticians very widely: it may be used for instance for the aggregate of artefacts coming off a particular production line, or for an infinite set of hypothetical coin-tosses. When we use statistics in the study of language, both the sample and the population are themselves linguistic items. One might wish, for instance, to make generalizations about the style of Gibbon's *Decline and Fall* without making word-counts for the whole of that massive text. If one makes inferences from randomly chosen passages of the text, then those passages are the samples and the entire text is the population. A linguist may use particular texts, or collect particular oral data, with a view to learning about a dialect or language as a whole. Here the whole language is the population: clearly it is not a closed or finite one, like *The Decline and Fall*, and sampling is the only, and not just the most convenient, method of studying it.

The first step in making an inference about a population from a sample is to calculate the descriptive statistics of the sample itself. Inferential statistics is built upon descriptive statistics. There is indeed a technical use of the word 'statistics', within the discipline which is popularly so called, to refer to the measurements of samples, such as means and standard deviations. The corresponding magnitudes in the population as a whole are known as *parameters*. In inferential statistics we are inferring from the statistics of samples to the corresponding parameters of the population.[1]

In a wholly uniform population it would be sufficient to take a single member in order to generalize about the population as a whole. But few populations consist entirely of members which are as alike as two peas: and even a population of peas would be full of minute variations in shape, size and weight. In a population containing variation—and of course all literary populations worth studying contain considerable variation—we need to devise methods of generalization despite variation between one sample and another.

If we wish to estimate a population parameter from a sample statistic there are three factors which always need to be taken into account. The first is the method by which the sample is chosen, the second is the size of the sample, and the third is the degree of confidence which can be placed in the estimate. If a sample is to be fairly representative of a population it must be a random sample, that is a sample in which each member of the population has an equal chance

---

[1] It is customary in statistical symbolism to distinguish between statistics and parameters by the use of Latin and Greek letters. The mean and standard deviation of a sample are given by the Latin symbols $\bar{x}$ and s; for the corresponding population parameters we use instead μ and σ.

of appearing. The choice of a random sample is not as simple a matter as might appear, particularly in a literary context; but discussion of the problems raised will be postponed until later in the book. The other two features will now be considered together. They are related since the degree of confidence which can be placed in an estimate depends among other things upon the number of items in the sample on which it is based.

Let us illustrate the relationship in a case in which we wish to estimate a population proportion from a sample proportion. Suppose we wish to find out what proportion of written English is made up of occurrences of the vowels 'a e i o u'. Suppose further that by some suitable process the first 100 letters of the previous paragraph have been chosen as a random sample on which to base this estimate. We find by counting that 41 of the 100 letters are tokens of one or other of the five vowels. Can we conclude that 41% of written English consists of uses of the vowels?

Obviously it would be foolish to regard the sample proportion as more than an approximation to the population parameter. We would not expect other samples to show exactly the same proportion. Indeed, if we count the next 100 letters we find that this time the number of vowels is only 34. We might take half-a-dozen samples each with different proportions. How closely can we expect the proportion in any given sample to approximate the proportion in the population as a whole?

The reliability of the estimate is found to depend on two things: the number of items in the sample and the numerical value of the proportion. It seems intuitively obvious that a larger sample will provide a better estimate of a population than a smaller one. But it turns out that the reliability of a sample does not increase in direct proportion to its size: an estimate of 0.375 based on our 200-letter sample would not be twice as reliable as an estimate of 0.410 based on our first 100-letter sample. To double the accuracy of our first estimate we would have to increase the sample fourfold: for it has been ascertained that the reliability of a sample increases in proportion to the *square root* of the increase in the number of items.

The numerical value of the proportion affects the reliability of the estimate in this way: it has been discovered that the reliability is greater if the proportions in the sample are very different from each other (one of the attributes being compared being close to 100% of the sample and the other being close to zero), and becomes less as the proportions approach equality at 50%. More precisely, it is found that the degree of error in the estimate varies directly as the square root of the product of the two proportions. Thus a sample containing items of two categories in the proportion 50/50 will be only 4/5ths as reliable an indicator as a sample of the same size containing items in the proportion 80/20; for the product of 50/100 and 50/100 is 2500/10000, whose square root is 50/100 whereas the square root of 1600/10000, the product of 80/100 and 20/100, is only 40/100.

Statisticians indicate the reliability of an estimate based on a sample by attaching to it a measure of the likely degree of error in the estimate: a measure which is known as the *standard error* of the estimate. The formula for the standard error of a proportion is

$$SE = \sqrt{\frac{pq}{N}}$$

where p represents the proportion of one of the categories in the sample, and q ($= 1 - p$) the proportion or frequency of the other, and N represents the total number of items in the sample. The formula takes into account the two factors mentioned above as affecting the reliability of the estimate of a proportion. The standard error varies inversely to the square root of the number of items in the sample, and directly as the square root of the product of the two proportions.

We can use this formula to calculate the standard errors of the two sample proportions ascertained above. For the first the value is:

$$SE = \sqrt{\frac{0.41 \times 0.59}{100}} = \sqrt{\frac{0.242}{100}} = 0.049.$$

For the second we get the value:

$$SE = \sqrt{\frac{0.34 \times 0.66}{100}} = \sqrt{\frac{0.224}{100}} = 0.047.$$

The value of the standard error of the proportion can range from 0 (when 100% of the sample is of the same kind) to 0.354, i.e.

$$\sqrt{\frac{0.5 \times 0.5}{2}}$$

(the value calculated for the least reliable of all samples, a single pair of items containing one item from each category).

If we consider words not letters in a text we will deal with proportions that are very distant from 50–50 ones. Very few words in any language occur at a rate of more than 5%. Correspondingly, for a given sample size, the standard error will be lower. For a 100-word sample, and a frequency of 5%, the standard error will be

$$\sqrt{\frac{0.05 \times 0.95}{100}} = \sqrt{\frac{0.0475}{100}} = 0.022.$$

If the frequency is 1% the standard error, for a 100 word sample, is 0.99%. It will be seen that as the frequency in the sample gets lower, the standard error is reduced absolutely, but becomes greater in proportion to the frequency itself.

Where the frequency is low (say, less than 1%) the formula for the standard error can be simplified without serious loss of accuracy. For if p is 0.01 or less, q is very near to unity, and therefore multiplying p by it makes very little difference to p. In that case the standard error can be calculated by dividing the square root of the frequency by the square root of the number of items in the sample. For a frequency of 1%, the standard error thus calculated would be

$$\sqrt{0.01/100} = 0.1/10 = 0.01$$

or 1%.

Unless we are given the standard error of an estimate, it is virtually worthless to be told the proportion of categories in a sample. We are given no clue to how likely it is that the population proportion will resemble at all closely the sample proportion. But once we have been given the standard error, there are a number of ways in which we can use it in estimating the population proportion or in making decisions about further sampling. These will be explained in the following section.

Exercise 13.

Here is a passage from an English translation of St Augustine's *Confessions*. Using it as a sample, give an estimate, with standard error, of the following parameters for the work as a whole:

(a) The proportion of words beginning with letters from the first half of the alphabet, up to M inclusive.

(b) The proportion of the text made up of personal pronouns.

> O Thou, my hope even from my youth, where wast thou, and whither wert thou gone from me? Was it not thou who madest me, and thou that didst distinguish me from the beasts of the earth and the fowls of the air? Thou madest me wiser than they, and yet I went walking through dark and slippery places, and I sought thee without myself, and I found not thee the God of my heart; but I came on out from thee even to the depths of the sea, and I distrusted and despaired of ever finding out the Truth.

Solution:

(a) $p = 0.5$ $SE = \sqrt{0.0025} = 0.05$.

(b) $p = 0.2$ $SE = \sqrt{0.0016} = 0.04$.

The great advantage of a standard error of an estimate is that it enables one to decide both how precise and how reliable it is as a guide to the population parameter. Precision and reliability vary inversely: we can use a sample proportion to make a very precise estimate of the population proportion, but if we do so the estimate will very likely not be right; or we can make a broad estimate of the population proportion with a high degree of confidence in it being correct. It is the standard error which tells us the degree of scope we have in this trade-off between precision and reliability.

The most precise estimate we can make of the population proportion is the sample proportion itself: we can make the estimate that the proportion in the population is exactly the same as it is in the sample. A precise estimate of this kind is known as a *point* estimate: it is an estimate assigning just a single value, a single point on the scale, to the population parameter. The precision is paid for by a considerable risk of error, and it is the standard error statistic which tells us the exact extent of this risk.

There is a body of statistical theory which enables one to calculate how far from the truth one's estimate of a population parameter is, on the basis of the standard error of the estimate. Some of the theory underlying such calculation will be given later in the book: for the moment it will suffice to explain some of its results. It can be shown, for instance, that if we have a random sample with a given proportion *p*, then of the possible populations from which it could have been drawn approximately 68% will have proportions lying no further than one standard error away from *p* in either direction; approximately 95% will be distant from *p* by no more than ±2 standard errors; and more than 99% will lie within 3 standard errors of *p*.

This can be illustrated in the case of the proportion of written English made up of the five vowels. The proportion in our first sample was 0.41 and the standard error was, to two decimal places, 0.05. We can use 0.41 as a point estimate of the population proportion and say that there is a chance of 95 out of 100 that the true population parameter is between 0.31 (which is 0.41 *minus* twice the standard error of 0.5) and 0.51 (which is 0.41 *plus* twice the S.E.). The standard error thus provides us with a means of associating how far out an estimate might be with the probability of its being thus far out. The odds are 20–1 against its being more than 10% out in either direction.

Similarly, for our second sample, the standard error to two decimal places was again 0.05; we can use this to say that there is a 95% chance that the population proportion is between 0.24 and 0.44. It will be seen that a large range of possible values for the population proportion is within the two-standard-error range of each of the two sample proportions.

We can, if we prefer, express out estimate of the population proportion in a different way. Instead of offering a single value plus an estimate of how far out it is likely to be, we can use the standard error to estimate the interval within which the population parameter is likely to fall. Thus on the basis of the first sample we can say, with a 95% confidence of being correct, that the population parameter falls between 0.31 and 0.51. This is called giving an *interval* estimate and the boundaries of the interval are called the *confidence limits* of the proportion. Since we have a 95% chance of being correct these are called 95% confidence limits. Such an interval estimate giving confidence limits does not offer any information which is. not already implicit in the giving of a standard error with a point estimate.

If we are not content with the 5% risk of error which is involved in using 95% confidence limits, we can reduce the risk, at the cost of precision, by widening the limits of the interval. We can use 99% confidence limits and say, with only one chance in a hundred of being wrong, that the population parameter as judged on the basis of our first sample falls between 0.26 and 0.56. We can do this because of the body of statistical theory referred to above which shows that more than 99% of the populations from which a sample could have been drawn will have parameters no further than three standard errors from the sample statistic.

The chances of being wrong here are in fact less than 1 in 100, since there was a degree of approximation in saying that 99% of the populations lie within the three-standard-error limit. Confidence limits for various proportions and various sizes of sample are given with a greater degree of accuracy in the following table.

95% and 99% confidence intervals for proportions.

Sample proportion	Size of sample															
	50				100				250				1000			
0.00	0	0	07	10	0	0	04	05	0	0	01	02	0	0	0	01
0.01					0	0	05	07	0	0	04	05	0	0	02	02
0.02	0	0	11	14	0	0	07	09	01	01	05	06	01	01	03	03
0.03					0	01	08	10	01	01	06	07	02	02	04	04
0.04	0	0	14	17	01	01	10	12	02	02	07	09	03	03	05	06
0.05					01	02	11	13	02	03	09	10	03	04	07	07
0.06	01	01	17	20	02	02	12	14	03	03	10	11	04	05	08	08
0.07					02	03	14	16	03	04	11	13	05	06	09	09
0.08	01	02	19	23	03	04	15	17	04	05	12	14	06	06	10	10
0.09					03	04	16	18	05	06	13	16	07	07	11	12
0.10	02	03	22	26	04	05	18	19	06	07	14	16	08	08	12	13
0.20	08	10	34	38	11	13	29	32	14	15	26	27	17	18	23	23
0.30	15	18	44	49	19	21	40	43	23	24	36	38	26	27	33	34
0.40	23	27	55	59	28	30	50	53	32	34	46	48	36	37	43	44
0.50	31	36	64	69	37	40	60	63	42	44	56	58	46	47	53	54

Derived from the table in Snedecor and Cochran, *Statistical Methods*, sixth edition, pp. 6–7.

Confidence intervals are given in per cent. For each size of sample four figures are given corresponding to each sample proportion. The outer pair of figures gives the 99% confidence limits, and the inner pair gives the 95% confidence limits.

Where the proportion is greater than 0.50, subtract from 1 and subtract each confidence limit from 100.

Suppose that we have a third 100-letter sample in which the number of vowels is 40. To find from the table the confidence limits for an estimate of the population proportion based on that sample, we look across the top line to the sample size of 100 and then look down until we reach the line for 0.40. We there find four numbers. The inner pair, 30 and 50, give us the 95% confidence limits in per cent: we can say, with 95% confidence, that the population proportion falls between 30 and 50%. The outer pair gives us the 99% confidence limits, 28 and 53%.

Note that the confidence limits make a statement about the ratio between the proportions in samples and the proportion in the population. What it amounts to is this: if the sampling is repeated over and over again, each time producing a new interval estimate, then in 95% of the cases the interval will cover the true population percentage. The assigning of confidence limits does not make any statement about the relationship between one sample and another.

By widening the intervals in our interval estimate we can, as we have seen, increase the likelihood of our estimate being correct from 95 to 99% and above. But there is no way of making an estimate that will be 100% correct, except the vacuous one that the population proportion will be between 0 and 1.0. For we cannot rule out the theoretical possibility of a totally unrepresentative sample with a proportion very different from the population proportion. The odds against such a sample may be very high, but however high, they always remain finite and therefore capable of being defeated.

Exercise 14.

On the basis of the following sample of 250 letters, use the table of confidence intervals to make a 95% confidence interval estimate of the relative frequency in written English of the letters 's', 'h', 'i'.

> There are very many to whose mind the cat cannot effect an entrance unaccompanied by 'harmless necessary'; nay, in the absence of the cat, 'harmless' still brings 'necessary' in its train; and all would be well if the thing stopped at the mind, but it issues by way of the tongue, which is bad, or of the pen, which is worse. (Fowler, *Modern English Usage.*)

Solution:

s = 0.08; confidence limits 0.05 to 0.12.
h = 0.02; confidence limits 0.01 to 0.05.
i = 0.06; confidence limits 0.03 to 0.10.

We have seen how the standard error can be used as an indicator of the reliability of an estimate from a sample of a given size. It can also be used in the opposite way, to determine how large a sample must be in order to achieve a specified degree of reliability. Suppose that we wish to use a sample to determine the frequency of individual words in a larger population (say, a book, or the work of a particular author). How large should the sample be in order to bring the probable error of the estimate within acceptable limits?

The probable error of an estimate is given by its confidence limits: if we give the 95% confidence limits, we are saying that there is a probability of 20–1 that the error of the estimate will not be greater than the distance between the point estimate and the limits. Consequently, the 0.95 probable error is roughly twice the standard error.[1]

Suppose, then, that we want our sample to estimate the frequency of individual words in the population with a 0.95 probable error. How large must we make the sample?

Since the 0.95 probable error is twice the standard error we have, from the definition formula of the standard error of a proportion

$$E = 2 \sqrt{pq/n},$$

where E = probable error.
Elementary algebra transforms this into

$$n = 4 \, pq/E^2.$$

We know that no word has a frequency of more than 0.1, and we know that the error of an estimate diminishes as the proportions approach 0 or 1. Hence we will be on the safe side if we substitute '0.1' for p, and '0.9' for q. E, we have decided, is to be 0.5%. Substituting in the expression above

$$n = 4 \times 0.1 \times 0.9/0.000025 = 14,400.$$

A sample of 14,400 or more words, then, will enable us to estimate the rate of words in a larger population with a probable error (at the 95% level of confidence) of not greater than 0.5%. A sample to produce the same degree of accuracy of estimate at the 99% level of confidence

---

[1] I use the term 'probable error' as it is used, e.g. by Caulcott (*Significance Testing*, 52). Other statisticians use it with a different technical meaning.

would have to contain approximately 23,900 words. If we want a sample which will estimate with similar accuracy *any* proportion (and not just the near-zero proportions of words) we need, for a 95% probable error of 0.005 a sample of 40,000 words. For the highest value that pq can take is 0.25 when $p = q = 0.5$. In that case the formula for n becomes $1/E^2$.

These results may appear discouraging. Only a large sample, it appears, will justify the drawing of conclusions about word frequencies with an acceptable degree of accuracy and confidence. This is only partly true. The large size of the sample for the 0.95 probable error of 0.5% is due to the stipulation that there should be a single sample permitting frequency estimate for all words including those with a very high rate of frequency (up to the unrealistic limit of 10% of the total). The great majority of words in most languages occur with frequencies of less than 1%. The sample size for a 0.95 probable error of 0.5% calculated as above for words occurring at a rate of 1% is 1584 words, and for words occurring at a rate of 0.5% it is 796. On the other hand, a probable error of 0.5%, which is perfectly acceptable if one is studying words occurring at a frequency of around 5%, is embarrassingly large in relationship to a frequency rate which is itself only 0.5%. The effect of accepting such a probable error is that genuine differences between frequencies in the underlying population will be masked by differences in the samples due to sampling variation. To reduce the probable error to 0.2%, at the 95% confidence level, for a word occurring with a frequency of 0.5%, we need a sample size of approximately 5000 words.

On the other hand it may seem surprising that in calculating the likelihood of error in estimating a population from a sample we have taken into account only the size of the sample, and nothing has been said about the size of the population. Surely a larger sample is necessary to make an estimate of a parameter in a population of 10,000,000 than to estimate one in a population of 10,000. Surely a sample of *Decline and Fall* which would be large enough to permit a reliable estimate of that work, or even of Gibbon's entire output, would be inadequate to base a generalization about the English language as a whole. The fact is that there is little difference between the size of sample to make an estimate of given reliability about a population of

10,000,000 or a population of 10,000, provided that the population is homogenous. If a sample from *Decline and Fall* would be a rash basis for generalization about the language as a whole, that is not because the language is a particularly vast totality, but because we have every reason to believe that it is not a homogenous totality. We do not need a larger sample to take account of *sampling variation* in a larger population; but we may very well need one to take account of variation due to other factors, such as conscious choice of authors or chronological development of the language.

There is one circumstance where we have to take into account the size of the population in calculating the reliability of an estimate, and that is not where the population is large in proportion to the sample but where it is small in proportion. If the sample constitutes a large part of the population, the normal method of calculating standard errors will underestimate the reliability of the estimate. In such cases we need to apply a correction to the standard error: having calculated it in the normal way we must then multiply it by

$$\sqrt{\frac{N - n}{N - 1}}$$

where N is the number of items in the population and n the number of items in the sample. If, for instance, we are studying a text 1001 words long, and we sample 511 words of it, we would have to multiply the standard errors we obtained by

$$\sqrt{\frac{1001 - 511}{1000}} = \sqrt{0.49} = 0.7.$$

Thus the effect of allowing for the size of the population is to reduce the standard error and accordingly to indicate greater reliability in the estimate.

Exercise 15.

You have reason to believe that the proportion of 'e's in written English is around 12%. How many letters would a sample contain which would enable you to verify this with a 95% probable error of 1%?

Solution: 4224.

As was said above, the assigning of confidence limits to an estimate based on a sample does not by itself tell us anything about the relationship between that sample and any other sample. We may, however, wish to compare samples. If two samples exhibit different proportions, we may wonder whether the difference is sufficiently small to be simply the result of sampling variation, or whether it is so large as to be statistically significant—that is to say, so large as to suggest that the two samples have been drawn from different populations. The standard errors of the estimates can be used so as to test the differences for statistical significance.

We can calculate a standard error for the difference between two proportions on the basis of the standard error of each proportion. On certain assumptions, which will be made explicit in the final chapters of this book, the simplest way to calculate standard error of the difference between two proportions is to take the square root of the sum of the squares of the individual standard errors. In a formula, where we have standard errors SE_1 and SE_2

$$SE \text{ of difference} = \sqrt{SE_1^2 + SE_2^2}.$$

Thus, if we return to our two original samples to estimate the number of vowels in written English, it will be remembered that the standard error of each estimate, to two decimal places, was 0.05. The square of this is 0.0025, and so the standard error of the difference between the two proportions is the root of 0.005, which is 0.071, to two decimal places.

This standard error in its turn can be used to measure the likelihood of the difference between the proportions occurring by chance. The procedure is parallel to the one adopted above in evaluating the accuracy of an estimate. Statistical theory can show that where there are differences between samples arising solely from chance sampling variation, the differences in approximately 68% of the cases will be less than one standard error, and in approximately 95% of the cases will be less than two standard errors. Consequently a difference of over two standard errors will have only a 1 in 20 chance of being due to mere sampling variation. On the other hand, a difference less than one standard error has a high probability of being due to chance. The common practice of statisticians working in social sciences or humanities is to take the 95% mark as the threshold of statistical significant: so that a difference is regarded as significant if the odds in favour of its happening by chance are no more than 1 in 20.

It is possible to calculate the probability of occurrence of any given deviation measured in terms of standard error units. It is difficult, however, to express the calculations in any very simple single formula. Consequently it is more convenient to find the probability by consulting a table. The table on p. 103 gives the probabilities of deviations of different sizes, measured in standard-error units.

To use it we first work out the actual difference between the proportions in two samples and express this difference in standard error units, to two decimal places. The probability of a difference as large as this occurring by pure chance will be found at the intersection of the appropriate row (corresponding to the first decimal place) and the appropriate column (corresponding to the second decimal place) on the table. Thus the difference between the proportion of vowels in our two samples was $0.41 - 0.34 = 0.07$. The standard error was itself 0.07, so the difference was one standard deviation. We look up 1.0 in the left most column of the table and then move along to the next row, the row under 0 for the second decimal place. We find the probability of 31.74%. We see that a deviation such as the observed one would occur by chance in nearly one in three of all pairs of samples. Since 31.74 is well above the 5% threshold we conclude that the difference between the two samples is due merely to chance and that they may very well be samples from a single uniform population.

We see from this table that a difference of 1.96 standard deviations gives a probability of exactly 5.0. This is the exact value to which we have hitherto been approximating when we said that 95% of samples would be within two standard deviations of the population norm, and vice versa. Similarly, the exact value for the 1% probability (which corresponds to the 99% confidence limits) is between 2.57 and 2.58. The probability of a difference as great as three standard errors is only 0.27%.

The table can be used to evaluate the probability not only of a difference between two sample proportions, but also the probability

TABLE 1. *The Probability of Occurrence of Standard-Error-Unit Deviations*

AD/SE	0.00	0.01	0.02	0.03	0.04	0.05	0.06	0.07	0.08	0.09
0.0	100.0	99.20	98.40	97.60	96.80	96.02	95.22	94.42	93.62	92.82
0.1	92.04	91.24	90.44	89.66	88.86	88.08	87.28	86.50	85.72	84.94
0.2	84.14	83.36	82.58	81.80	81.04	80.36	79.48	78.72	77.94	77.18
0.3	76.42	75.66	74.90	74.14	73.38	72.64	71.88	71.14	70.40	69.66
0.4	68.92	68.18	67.44	66.72	66.00	65.28	64.56	63.84	63.12	62.42
0.5	61.70	61.00	60.30	59.62	58.92	58.24	57.54	56.86	56.20	55.52
0.6	54.86	54.18	53.52	52.86	52.22	51.56	50.92	50.28	49.66	49.02
0.7	48.40	47.78	47.16	46.54	45.92	45.32	44.72	44.12	43.54	42.96
0.8	42.38	41.80	41.22	40.66	40.10	39.54	38.98	38.44	37.88	37.34
0.9	36.82	36.28	35.76	35.24	34.72	34.20	33.70	33.20	32.70	32.22
1.0	31.74	31.24	30.78	30.30	29.84	29.38	28.92	28.46	28.02	27.58
1.1	27.14	26.70	26.28	25.86	25.42	25.02	24.60	24.20	23.80	23.40
1.2	23.02	22.62	22.24	21.86	21.50	21.12	20.76	20.40	20.06	19.70
1.3	19.36	19.02	18.68	18.36	18.02	17.70	17.38	17.06	16.76	16.46
1.4	16.16	15.86	15.56	15.28	14.98	14.70	14.42	14.16	13.88	13.62
1.5	13.36	13.10	12.86	12.60	12.34	12.12	11.88	11.64	11.42	11.18
1.6	10.96	10.74	10.52	10.32	10.10	9.90	9.70	9.50	9.30	9.10
1.7	8.92	8.72	8.55	8.36	8.18	8.02	7.84	7.68	7.50	7.34
1.8	7.18	7.02	6.88	6.72	6.58	6.44	6.28	6.14	6.02	5.88
1.9	5.74	5.62	5.48	5.36	5.24	5.12	5.00	4.88	4.78	4.66
2.0	4.56	4.44	4.34	4.24	4.14	4.04	3.94	3.84	3.76	3.66
2.1	3.53	3.48	3.40	3.32	3.24	3.16	3.08	3.00	2.92	2.86
2.2	2.78	2.72	2.64	2.58	2.50	2.44	2.38	2.32	2.26	2.20
2.3	2.14	2.08	2.04	1.98	1.92	1.88	1.82	1.78	1.74	1.68
2.4	1.64	1.60	1.56	1.50	1.46	1.42	1.38	1.36	1.32	1.28
2.5	1.24	1.20	1.18	1.14	1.10	1.08	1.04	1.02	0.98	0.96
2.6	0.94	0.90	0.88	0.86	0.82	0.80	0.78	0.76	0.74	0.72
2.7	0.70	0.68	0.66	0.64	0.62	0.60	0.58	0.56	0.54	0.52
2.8	0.52	0.50	0.48	0.46	0.46	0.44	0.42	0.42	0.40	0.38
2.9	0.38	0.36	0.36	0.34	0.32	0.32	0.30	0.30	0.28	0.28
3.0	0.27	0.26	0.26	0.24	0.24	0.22	0.22	0.22	0.20	0.20
3.1	0.20	0.18	0.18	0.18	0.16	0.16	0.16	0.16	0.14	0.14

of a given estimate of the population being in error by an amount expressed in standard error units. It can be used also in conjunction with any kind of standard errors of estimate, not just the standard errors of proportions discussed in this chapter, but also the standard errors of other sample statistics to be explained later.

The relationships between various samples, and the likelihood of their being samples drawn from the same population, can be rep-

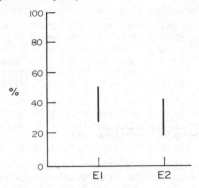

resented graphically. Along the base of the graph we plot the various samples and along the ordinate we plot the population proportion. The interval estimate of the population proportion given by each sample is represented by a vertical line whose end-points represent the upper and lower confidence limits.

The two lines on the graph above represent the interval estimates for the proportion of vowels in English based on the two samples above. It will be seen that they overlap considerably.

8

Testing for Significance

AT THE end of the previous chapter we introduced a method of assessing the statistical significance of the difference between proportions observed in two samples. In the present chapter the general strategy of significance testing will be explained and the most universally useful test of statistical significance will be described.

The test of statistical significance in the previous chapter was presented in the context of a question whether two samples with given proportions could have come from the same population. There are many other types of question which tests of statistical significance can be used to answer. Perhaps the samples concern magnitudes and not frequencies, and we wish to know whether two samples with differing means can be regarded as coming from a single population. Or perhaps we have a population where the mean of a certain magnitude, or the frequency of a certain attribute, is known, and we wish to know whether a sample with a particular mean or proportion could have been drawn from that population. In all such cases we observe a deviation between two statistics, or between a statistic and a parameter, and we test the deviation to see whether it is too large to be accounted for by the ordinary variation between sample and sample which statisticians call 'sampling error'.

If we obtained a pair of dice and rolled them once and both dice turned up sixes, we would think nothing of it. If the first five rolls are double sixes, we would be surprised and puzzled. If we continue rolling 20 times and each time a pair of sixes appear, we would very likely be convinced that the dice were loaded. Testing a pair of dice in this way may be regarded as a type of significance testing.' We are comparing the proportion of double sixes to be found in our sample of 20 throws with the proportion to be expected in the unlimited population of throws of a pair of fair dice—a proportion which we can calculate *a priori* since a fair die is by definition one in which each of the six faces has an equal chance of appearing. When we ask whether this pair of dice is fair, we are asking whether our 20 throws can be a sample from the population of throws with fair dice.

In literary contexts significance testing can be useful in connection with authorship attribution. If we have a work which is doubtfully attributed to an author, we may make a comparison between quantifiable features of the dubious work and the same features of the established corpus. We can calculate the deviations between the statistics and ask whether they are too great to occur as the result of sampling variation. If they are, then they mark genuine differences of style, which call for an explanation. Diversity of authorship is not of course the only possible explanation: a deliberate variation of style by one and the same author might be the explanation. This is a matter for the literary scholar, not the statistician, to decide. But the discovery of significant differences may shift the onus of explanation from the denial to the affirmation of authenticity. On the other hand, if the differences are not statistically significant, they cannot be used as the basis of a stylistic argument for spuriousness. Thus a test of statistical significance may form part, though it can never form the whole, of the solution of an authorship attribution problem.

No matter what its subject matter, a significance test can never establish any proposition with absolute certainty. Inferential statistics deal not with certainties but with probabilities. No matter how great the deviation between two statistics may be, there is always at least a minute probability that it is the result of chance. A significance test can only establish how probable it is that the deviation is due to sampling error. But this is no great handicap. We can decide for ourselves how small a probability has to be before we regard it as putting an end to an argument. We may decide that if the probability of a divergence as great as that observed between our statistics is less than 1 in 20 (0.05), then we will regard the two samples as coming from different populations. This is called 'testing at the 0.05 level'. Or we may decide to be more cautious about accepting a result as significant and refuse to do so unless the chances of it occurring as a result of sampling error are less than 1 in 100. That will be 'testing at the 0.01 level'. The results of the tests will then be described, respectively, as 'significant at the 0.05 level' or 'significant at the 0.01 level'.

The first step in significance testing is to formulate a hypothesis. The hypothesis in a significance test is about the value of a population parameter. We may surmise that the parameter in a population is twice as large as, or half the size of or exactly the same as, the parameter in another population. A specific hypothesis about the relation between two parameters is called the null hypothesis and it plays a special part in significance testing. Let us consider as an example of a null hypothesis the hypothesis that two parameters are the same. Since there are many different ways in which two parameters can differ, there are many different alternative hypotheses to this null hypothesis. These hypotheses are not tested directly: the way in which an investigator discovers evidence in favour of them is by collecting facts to refute the null hypothesis.

If we are testing whether a sample could have been drawn from a population with a known parameter, then the null hypothesis is that the population from which the sample is drawn has a parameter equal to that of the known population parameter. If we are testing whether two samples, A and B, could have been drawn from the same population, we are testing whether the parameter of the population from which A is drawn is no different from the parameter of the population from which B is drawn. We are not, of course, testing to see whether the sample statistic itself is the same as the parameter of the population: the whole essence of sampling variation is that these two will, about ninety five per cent of the time, be different, but the difference will not be significant. To regard a divergence discovered between two statistics as significant is the very same thing as to reject the null hypothesis about the underlying population parameters.

Thus, if we are testing two texts in the course of trying to decide whether they are by the same author, we will find various differences between word-frequencies, word- and sentence-length and the like. We set up the null hypothesis that there is no difference between these parameters in the populations from which the passages are drawn and that the variations which are observed are due simply to sampling error. We decide the level of significance at which we are to test. If the chance probability of the observed divergences is less than the level of probability fixed in advance, we reject the null hypothesis.

The level of probability which is fixed in advance is designated by the greek letter *alpha* (α). As was mentioned above, the alphas usually chosen are 0.01 and 0.05. The choice between these two is made by the investigator. The choice is not arbitrary: it is made partly on the basis of the subject matter to be investigated and partly on account of the purpose of the investigation.

In the physical sciences, where exact measurement is possible and experimental variation is comparatively small, stringent testing with αs as low as 0.001 is common. In the social sciences, where exact measurement is often delusory and individual situations display great variation, the 0.05 level of testing is probably the most frequent. Literary studies stand somewhere between. On the one hand literary phenomena, when they are quantifiable, can be measured with an exactitude comparable to that of the physical sciences. On the other hand, the variability from individual to individual in literary matters is as great as any variability to be found in sociology and economics: the factor of human choice is ever present to confound the exceptionless generalization. But the statistical investigator of literary phenomena has one great advantage over his colleagues in either the exact or the social sciences: he can study large samples at comparatively little cost. The quantitative study of texts, even when performed by computer, is much cheaper than extensive experimentation in the natural sciences or large-scale surveys in the social sciences. Since the reliability of statistical results is related to sample size, this factor is relevant to the setting of the appropriate alpha for testing. In most literary studies an alpha of 0.05 or 0.01 seems to be chosen, and the choice between the two may be guided by a consideration of what type of mistake about hypotheses would be more damaging to the investigation.

When we decide whether to accept or reject the null hypothesis, two kinds of mistake can be made. We may reject the null hypothesis when it is true and we may accept it when it is false. The first error is an error of credulity: it is taking seriously something which is a mere coincidence. Statisticians call it a type I error. The second error is an error of scepticism: it is treating as a matter of luck something which genuinely calls for explanation. Statisticians call it a type II error.

Pascal's famous wager about the existence of God can be thought of as an early reflection upon the difference between type I and type II errors. If we set up the null hypothesis that God does not exist, to believe that there is a God when there is not would be a type I error: if there is no God, the null hypothesis is true and the theist errs by credulity. To fail to believe that there is a God when there is would be a type II error: if there is a God, then the atheist errs by his scepticism. Pascal by arguing that an error of the second kind was much more disastrous in its consequences than an error of the first drew the conclusion that one should believe in God. It is likewise by reflecting on the consequences of the two types of mistake that statisticians decide on the appropriate alpha for a statistical test.

The situation can be represented graphically thus

	Null hypothesis true	Null hypothesis false
Null hypothesis accepted	Correct	Type II error
Null hypothesis rejected	Type I error	Correct.

The probability of making a type I error is alpha: it is the probability which we have decided to accept as the threshold of statistical significance. But the probability of making a type II error is not known. We do know that if we increase the probability of making a type I error we decrease the possibility of a type II error; but that is not the only factor to be taken into account. The likelihood of a type II error is greater the closer the true hypothesis is to the null hypothesis: if the null hypothesis is that two parameters are unequal, but only very slightly, then the danger of a type II error is great.

Whether this is so, however, is not something that is under the investigator's control: so his decision has to be based on the relative unattractiveness of the two kinds of error. If one is attempting to prove the inauthenticity of a work widely attributed to an author, it will clearly be important to make one's case as strong as possible and to disregard any evidence which might be attributed by critics to mere coincidence: a 0.01 α under a null hypothesis of authenticity will therefore be more suitable than a 0.05 one. If, on the other hand, a court is uncertain about the genuineness of a confession attributed to an accused, and calls in a literary statistician as an expert witness, a significance test with an α of 0.05 might be sufficient to cast a reasonable doubt on the evidential value of the confession. The decision about the appropriate α is closely connected with the question of the burden of proof, whether scholarly or forensic.

Once the appropriate level of significance has been settled, the statistician proceeds to calculate the statistics by which he wishes to test his hypothesis. He calculates the divergence between the statistics worked out from his data and the population parameter as if the null hypothesis is true. He then evaluates the probability of the observed divergence. If the probability is less than α, he rejects the null hypothesis. If the probability is greater than α, he can choose between two alternatives. If it is considerably greater than α (say, a probability of 0.3 with an α of 0.05) he will be well advised to accept the null hypothesis. If it is only just greater than α (say, a probability of 0.02 with an α of 0.01) he will probably not accept the null hypothesis, but instead to repeat the test with a larger sample in the hope of achieving a more definitive result.

The procedure of significance testing outlined above applies to the testing of statistics of many different kinds—means, medians, variances, proportions and so on. In previous chapter we saw how the procedure worked in the case of testing the significance of the difference between two proportions. In that case, the probability of the divergence was evaluated in three stages: the standard error (SE) of the difference was worked out, the divergence was transformed into SE units, and the probability of the thus standardized divergence was ascertained from a table. The same steps are involved in the procedures for ascertaining the significance of the difference between two means or medians or variances, except that the appropriate SE formula varies from statistic to statistic, as will be explained later. There is, however, another kind of significance test which does not involve the calculation of a SE, a test which is known as a chi-square (χ^2) test. Since the χ^2 test is the significance test of the widest and most general application, the rest of the chapter will be devoted to explaining the method of its operation.

The calculation of χ^2 is best explained by working through an example in detail. It has been suggested that among Greek authors there is a great deal of individual variation in the choice of the syntactic category of the final word of a sentence, and that a study of the last words of sentences can provide a test of authorship in Greek prose. To test this we would need to study a large number of samples from different authors, classify the last words of sentences into different categories, and see whether the differences in frequency between the categories were (a) significant between one author and another and (b) insignificant between two texts of the same author. A simple way to begin would be to classify all words into three categories: nouns, verbs and other words. We can then compare pairs or groups of texts on the basis of the way in which the final words of sentences fit into this classification.

For instance, in the Aristotelian corpus there are two treatises on rhetoric. One, in three books, known as the *Rhetoric*, is generally held by scholars to be genuine. The other, a shorter text known as the *Rhetoric to Alexander*, is commonly regarded as spurious. Let us make a comparison between a sample from each of them: the first 100 sentences of *Rhetoric A* and the first 100 sentences of the *Rhetoric to Alexander*. We first classify each sentence in each sample into one of three categories: sentence ending with a noun, sentence ending with a verb, sentence ending with some other kind of word. We can construct the following table. (A table of this kind is called by statisticians a contingency table.)

	Last words of sentences			
Work	Noun	Verb	Other	Total
Rhetoric A ss. 1-100	28	32	40	100
Rhetoric to Alexander ss. 1-100	27	52	21	100
Totals	55	84	61	200

The contingency table shows that there are similarities and differences between the types of last words in the two samples. Verbs end sentences much more often in the *Rhetoric to Alexander* than they do in the *Rhetoric*, and many more last words in the latter belong to categories other than noun or verb. On the other hand, nouns are used to end sentences with approximately the same frequency in each sample. How are we to know whether the differences between the two samples are due merely to sampling variation or are statistically significant?

It is here that the χ^2 test is applied to test the null hypothesis that the two samples come from the same population. The principle of the application of the χ^2 test to such a case is very simple. If the null hypothesis is true, then not only the *Rhetoric* sample, and not only the *Rhetoric to Alexander* sample, but also the sample formed by putting these two together, is a sample from one and the same population. But as the total sample is larger than either of its two parts, it will provide a better estimate of the underlying population parameters. We therefore use this larger sample as an indicator of the population, derive from it an expected value for each category in each of the smaller samples, and measure the size of the differences between these expected values and the values actually observed.

The calculation of the expected values is a simple matter. The total sample contains 84 verbs. If the two small samples were from the same population, we would expect these verbs to be distributed between them in proportion to their length. Since, in fact, the two samples are of equal length we would expect the number of verbs in each sample to be 42. So the expected value in each of the two cells in the column 'verb' will be 42. We can work out the other columns similarly. In general, we work out the expected value for each cell in the following way:

$$\text{expected value} = \frac{\begin{array}{c}\text{total of row in which cell occurs}\\ \times \text{ total of column in which cell occurs}\end{array}}{\text{grand total of all items}}.$$

In the case of the expected value of verbs in *Rhetoric* we get, using this formula

$$\text{expected value} = \frac{84 \times 100}{200}.$$

We can set out the expected values in a table of the same form as our original contingency table, thus:

Last words of sentences			
	Noun	Verb	Other	Total
Rhet. sample	27.5	42	30.5	100
R. to Al. sample	27.5	42	30.5	100
Totals	55	84	61	200

The calculation of expected values in this example is particularly easy because there are only two samples, each of equal size, and each with a conveniently round number of items. This will not always be the case. In particular, if the two samples are not of equal size the expected values in the two cells of each column will not be the same as each other.

Having obtained expected and observed values for each cell, we are in a position to calculate the chi-square statistic. To calculate χ^2 we square, for each category, the difference between the observed and expected value, and divide the squared difference by the expected value. We then sum the quotients thus obtained, and the resulting sum is the χ^2 for the table. It will always have a positive value because of the squaring of the differences. The calculation can be expressed in a formula thus

$$\chi^2 = \sum \frac{(O - E)^2}{E}$$

where

O = observed frequency for a cell.
E = observed frequency for a cell.
Σ = instructs to sum over cells.

For our data in the tables above we can set out the calculation thus: (we start at the top left and evaluate column by column; but it is no longer necessary to keep the individual cells separately identified).

Calculation of χ^2 for above tables.

O	E	$O - E$	$(O - E)^2$	$(O - E)^2/E$
28	27.5	0.5	0.25	0.01
27	27.5	−0.5	0.25	0.01
32	42	−10	100	2.38
52	42	10	100	2.38
40	30.5	9.5	90.25	2.95
21	30.5	−9.5	90.25	2.95
	Total χ^2			10.68

The χ^2, then, for the distribution of the three kinds of sentence-ending in our two samples is 10.68. The final stage of our significance testing is to evaluate the probability of a χ^2 of this size: so far we do not know whether a χ^2 of this value marks a small or large difference between our two samples.

As when evaluating the probability of a standardized difference between two proportions, we need to consult a table to answer the question. But there is a further complication in this case, since the significance of a χ^2 depends, among other things, on the size of the contingency table. The table we have used contains two rows and three columns; but χ^2 tests can be applied to contingency tables of any size, from 2 × 2 upwards. The larger the table, the higher the χ^2 is likely to be. For 3 × 2 contingency tables such as ours, the probability of obtaining a χ^2 equal to or larger than given values is shown in the following table:

χ^2	probability
1.386	0.50
4.605	0.10
5.991	0.05
9.210	0.01

From the table, we see that the probability of obtaining a χ^2 as high as 10.68 is less than 0.01. The difference between the two rhetorics in this respect is therefore statistically significant at the 0.01 level.

Tables of the probabilities associated with particular values of χ^2 are found in all statistical textbooks and one will be found at the end of this book. They are designed so as to be capable of use with contingency tables of any size, and for this reason the left-hand column of the table specifies the number of *degrees of freedom* involved. The degrees of freedom in a contingency table are the number of ways in which the observed values in the individual cells of the table can vary while leaving unchanged the characteristics of the overall sample represented by the table as a whole. Let us illustrate what this means by reference to the contingency table above.

There are two rows in the table, corresponding to the two rhetorics. (In counting the rows and columns of a contingency table, we do not count the lines giving the totals.) The total of nouns in the first row might have been anything between 0 and 55; but once the total of nouns in that row is fixed as n, then the total in the second is also automatically fixed as 55-n, since there are 55 nouns in the overall sample. Similarly, given the number of verbs in the first row, the number of verbs in the second row is fixed; and so also with the third category of words that are neither nouns nor verbs. So although we have two rows in the table, these two rows correspond to only one possibility of variation: for given the figure in the first row, the figure in the second row is fixed by the properties of the overall sample. We express this by saying that the two rows give only 1 DF.

If we turn from the rows to the three columns, here again we find that there are not three independent ways in which the numbers in the cells can vary. If the number of nouns and verbs among the last words is known for a sample, the number of words which are neither can be calculated by subtracting these numbers from the size of the sample as a whole; and in general, given the number of the total in any two cells, we can work out the number in the third cell as 100-n. The three columns give only 2 DF.

An analogous argument can be applied to a contingency table of any size. We find that if the table has r rows and c columns (an '$r \times c$ contingency table') then the DF can be calculated by the formula

$$DF = (r - 1) \times (c - 1).$$

In calculating the expected values for a contingency table it is not necessary to use the formula

$$\text{expected value} = \frac{\text{row total} \times \text{column total}}{\text{grand total of items}}$$

for every value, as we did in the explanation above. It is necessary to do so only for the number of values corresponding to the DF in a table; the remaining values can then be ascertained by subtraction. The DF belonging to a table can indeed be regarded as the number of independent calculations which are necessary to determine the expected values from the overall totals.

In evaluating the probability of a χ^2 from a contingency table, therefore, we enter the χ^2 table at the row corresponding to the DF $(r - 1)(c - 1)$. We then move along the row until we reach the value of χ^2 corresponding to the alpha we have determined. If the χ^2 derived from the contingency table is greater than this, then the null hypothesis is rejected.

It is important to appreciate that the values in the cells of the contingency table must be absolute frequencies and not proportions. If the samples contributing to the table are all of equal sizes, it will not, in general, matter whether occurrences or proportions are used. But if the samples are of different sizes, the larger samples should obviously make a greater contribution to determining the characteristics of the overall sample on which the expectations are to be based. This weighting will be lost if all the frequencies are reduced to proportions.

To illustrate the calculation of DF, and the operation of χ^2 when samples are of different sizes, let us work through another application of the last-words test to the works of Aristotle and the Aristotelian corpus. This time, instead of drawing up two separate tables for observed and expected values, we will put the expected values in brackets after each observed value.

The three samples are three books of Aristotle's Metaphysics, each in their entirety: *Metaphysics* α, *Metaphysics* E and *Metaphysics* Θ.

Work	Noun	Verb	Other	Total
Metaphysics α	17 (14.7)	30 (27.1)	23 (28.1)	70
Metaphysics E	16 (23.1)	48 (42.6)	46 (44.3)	110
Metaphysics Θ	68 (63.2)	108 (116.3)	124 (120.6)	300
Total	101	186	193	480

We can apply to this table the formula for χ^2

$$\chi^2 = \frac{(O - E)^2}{E}$$

$$= \frac{(2.3)^2}{14.7} + \frac{(-7.1)^2}{23.2} + \frac{(4.8)^2}{63.2} + \frac{(2.9)^2}{27.1} + \frac{(5.4)^2}{42.6} + \frac{(-8.3)^2}{116.3}$$

$$+ \frac{(-5.1)^2}{28.1} + \frac{(1.7)^2}{44.3} + \frac{(3.4)^2}{120.6}$$

$$= 0.36 + 2.17 + 0.36 + 0.31 + 0.68 + 0.59 + 0.93$$

$$+ 0.07 + 0.10$$

$$= 5.57.$$

Is a χ^2 of 5.57 significant? Let us suppose that we have fixed alpha for this test at 0.01. We work out the DF involved: with a table of three rows and three columns the number of DF = $(3 - 1) \times (3 - 1) = 4$. Entering the χ^2 table at the back of the book with 4 DF we discover that the value of χ^2 required for significance at the 1% level is 13.277. As our value is well below that we decide that the differences between the different books of the Metaphysics may well be chance variations due to SE. Even if we had set our α at 0.05 the differences would not have been declared significant, since at that level a χ^2 of 9.488 is required for significance. In fact a value such as the one we obtained could in the long run be expected to occur by chance more than 20% of the time.

Exercise 16.

The following contingency table shows the number of sentences ending in nouns, verbs or otherwise in samples, each of 100 words, from two of Aristotle's minor works. Use a χ^2 test to determine whether the differences between the two works are statistically significant.

Work	Noun	Verb	Other
De Somno	30	40	30
Do Insomniis	26	40	34

Solution: $\chi^2 = 30/56 = 0.54$ (not significant).

There are a number of conditions which must be satisfied if a χ^2 test is to be appropriate. The first is that all the items involved in the test must be independent of each other; that is to say, each item must appear in one and in no more than one of the cells, and no item must be explicitly or implicitly counted more than once. The second is that the number of items expected in each cell must reach a certain minimum. Consequently, if there are many categories involved in the classification, larger samples will be necessary. A commonly offered rule of thumb is that the expected frequency should be no less than five in at least 80% of the cells. This condition does not usually present any difficulty in literary studies. If a problem of this kind does arise, it is usually overcome by combining categories. For instance, a fuller study of the categories of sentence-endings might list nouns, verbs, adjectives, pronouns and other words. We would thus have five categories instead of the three we have been working with. If it turned out that the expected values for pronouns fell below five in a number of cases, we could combine pronouns with other words and use a four-column classification instead of the proposed five-column one.

The reader will have noticed in following the previous examples that the expected values in contingency tables include fractional values while the observed values are all whole numbers. χ^2 itself is a continuous function which can take any value, while the values whose divergences it measures are discrete and can take only whole number values. In most cases this does not matter, but if we have a contingency table with only two rows and two columns, with only a single DF, then the continuous χ^2 function will underestimate to a certain extent the actual probabilities of deviations. To allow for this we have to make an adjustment to the calculation of χ^2 which is known as Yates' correction. Instead of simply squaring the difference between the observed and expected values, we reduce the difference by 0.5 before squaring it. We subtract 0.5. from what statisticians call the *absolute difference* between O and E, the absolute difference being what we get by subtracting E from O and disregarding any minus sign which may occur in the result. The formula for calculating χ^2 with Yates' correction is

$$\chi^2 = \frac{(|O - E| - 0.5)^2}{E}$$

where $|O - E|$ stands for the absolute difference between O and E. We can illustrate the use of Yates' correction with an example. A sample of the nouns used by Bunyan and Macaulay, consisting of the nouns beginning with the letter N in a sample from each of them of approximately 16,000 words, yields the following distribution of nouns of Teutonic origin and nouns of Romance origin. In the contingency table the expected values are put in brackets after the observed ones. (See Herdan, p. 63.)

Author	Teutonic nouns	Romance nouns	Total
Bunyan	22 (17.3)	16 (20.7)	38
Macaulay	18 (22.7)	32 (27.3)	50
Total	40	48	88

The calculation of χ^2 from this table may be set out as follows:

$O - E$	$|O - E|$	-0.5 $(|O - E| - 0.5)^2$	$(O - E - 0.5)^2/E$
4.7	4.2	17.64	1.02
−4.7	4.2	17.64	0.78
−4.7	4.2	17.64	0.85
4.7	4.2	17.64	0.65
Total χ^2			3.20

If we enter the χ^2 table for one DF we find that the value 3.38 is not significant at the 0.05 level. Yates' correction here has meant the difference between significance and non-significance, for without it the value of χ^2 would be 3.96, which is above the critical value of 3.841. With a larger number of items, the importance of Yates' correction diminishes.

Exercise 17.

Use a χ^2 test to discover whether the differences recorded below are statistically significant.

Occurrences of 'To' followed by 'the' in two samples of Scott's *Antiquary*.

Sample	'To' followed by 'the'	'To' not so followed	Total
c. 14	10	90	100
c. 21	20	80	100
Total	30	170	200

Solution:
Without Yates' correction—$\chi^2 = 3.92$; just significant at 0.05 level.
With Yates' correction—$\chi^2 = 3.18$.

9

The Comparison of Means

IN A previous chapter we discussed the estimation of population proportions on the basis of sample proportions and explained how differences between sample proportions could be tested for statistical significance. Very similar techniques may be used in the estimation of other population parameters and in the assessment of the significance of other sample statistics. In the present chapter we shall concentrate on inferences from sample means to population means and comparisons between means occurring in different samples.

Just as we use a sample proportion as our best estimate of a population proportion, so we can use a sample mean as an estimate of a population mean. A sample mean will, of course, hardly ever be exactly the same as the mean of the population from which it is drawn: we must expect, with means as with proportions, that there will be some sampling error. But under conditions of simple random sampling, which means that every possible sample has an equal chance of being chosen, a sample mean will provide an unbiased estimate of the population mean: that is to say it is just as likely to overestimate it as to underestimate it. If we took a large number of samples the overestimates and the underestimates will tend to cancel each other out; if we calculate the mean of the sample means, this should come ever closer to the population mean the more samples we take. Sampling error in estimating the mean, like SE in estimating a proportion, diminishes in proportion to sample size. But in the case of estimating the mean there is another factor to be taken into account which affects the size of the SE. This error varies in proportion to the standard deviation (SD) of the population: that is to say, the greater the degree of variation we have between members of a

121

population, the greater spread there will be between means drawn from different samples.

These factors are taken into account in the formula for calculating the *standard error of the mean* (SEM). It will be recalled that a standard error is a measure of the sampling error to be expected when a particular sample statistic is used as an estimate of a population parameter. The SEM thus indicates the reliability of an estimate of a population mean based on a sample mean. It is equal to the SD of the population divided by the square root of the number of items in the sample. It is expressed in the following formula:

$$SEM = \frac{\sigma}{\sqrt{N}}$$

where σ is the population SD, and N is the number of items in the sample. How is this formula applied?

The IQ of children is defined in such a way that we know that the mean IQ of the population is 100 and the SD is 15. Consequently, if we take a sample of 36 children and test their IQs and then calculate the sample mean IQ we can obtain the SEM by substituting into the formula:

$$SEM = \frac{15}{\sqrt{36}} = 2.5.$$

However, since we already know that the population mean is 100, calculating the sample mean and its SE may well seem a futile exercise. It would be, if the only function of the SE were to assess the reliability of an estimate of an unknown parameter. It may also be used to assess the likelihood that a sample exhibiting a particular statistic is a random sample from a given population. In the case in point we could make use of the SEM to evaluate the hypothesis that the sample of 36 children was indeed a random sample, and that the children formed a typical group and were not, on average, more or less intelligent than the population at large.

Suppose, for instance, that the mean IQ of the sample was 112.5. We wish to test the null hypothesis that there is no difference between the mean of the population which this represents and the population as a whole, which is 100. We know from the formula above that the SEM is 3. The difference between the observed sample mean and the population mean is 12.5. We turn this difference into a z-score by dividing it by the SE. The z-score is 5; such a score was too great to be included in our table: in fact the probability of such a difference is only 0.0000573%. Hence, whether we decided to fix our α at 0.05 or 0.01 or even 0.001 we reject the null hypothesis, and conclude that we are dealing with a very untypical sample of children.

This simple example which we have considered is in several ways untypical. In most cases we do not know what the population SD is and we have to use the sample itself as the basis of an estimate of it. When, in an earlier chapter, we learnt how to calculate the SD of a distribution we used the formula

$$\sqrt{\frac{\Sigma x^2}{N}}$$

where

Σx^2 = the sum of the squared deviations;
N = the number of items in the distribution.

This formula is the appropriate one where the distribution in question embraces an entire population. When we wish to use the sample SD to estimate the population SD, matters become more complicated. We must distinguish between large and small samples. If the sample is a large one, containing say more than 30 items, then the sample SD calculated by the above formula is a good estimate of the population SD. If the sample is a small one, its SD thus calculated will provide a perceptibly biased estimate of the population SD always underestimating it, and a different formula must be used.

Let us consider first the simpler case of the large sample. We can use the SD of the sample in calculating the SE estimate of the mean according to the formula above, so that

$$\text{SEM} = \frac{\text{SD of sample}}{\sqrt{N}}.$$

We can then use this SE to calculate z-scores and confidence limits, as in the case of proportions. Let us illustrate in an example. The number of sentences in the Epistle to the Ephesians is 80; the mean length in words is 30.31 with a SD of 28.7. We calculate the SE from these figures as 28.7/8.94 = 3.21. We can then set approximate 95% confidence limits by adding two SEs in either direction, and say that the population from which that Epistle is drawn has a mean sentence length of between 23.89 and 36.71.

Like the procedure for setting confidence limits, the procedure for comparing two sample means and testing the significance of the difference between them is similar to the corresponding one for sample proportions. We first set the appropriate α; we set up the null hypothesis that there is no difference between the means in the two populations which the samples represent; we calculate a z value by dividing the observed difference by the appropriate SE of the difference and reject the null hypothesis if the z value is above the critical one corresponding to the selected α. The SE of the difference between the two means is given by the formula

$$\sqrt{\left(\frac{s_1^2}{n_1} + \frac{s_2^2}{n_2}\right)}.$$

Let us illustrate the procedure by comparing the mean sentence-length of the Epistle to the Ephesians with that of the Epistle to the Galatians. Galatians contains 166 sentences; it has a mean of 13.78 words and a SD of 9.92 words. The SEM of Galatians is therefore 9.92/12.8 = 0.77. Substituting in the formula above we calculate the SE of the difference between the two means as

$$\sqrt{\frac{9.92^2}{166} + \frac{28.7^2}{80}} = \sqrt{\frac{98.41}{166} + \frac{823.69}{80}} = \sqrt{0.59 + 10.30} = 3.30.$$

The observed difference between the means is 16.53; we divide this by the SE (3.3) to obtain the standardized normal deviate or z-score which is 5.00. This greatly exceeds the z-score (2.58) corresponding to an α of 0.01; we can say, therefore, testing at the 0.01 level, that the difference between the two means is highly significant and the null hypothesis should be rejected. The two Epistles cannot be regarded as random samples from the same population. (Which is not the same thing as saying they cannot be by the same author!)

If the samples are smaller and contain less than, say, 30 items, then the procedure is more complicated in two ways. In the first place, the calculation of the SD must be corrected to remedy the bias noted above. In the second place, the probabilities associated with particular divergences are different.

Instead of the formula $\sqrt{\Sigma x^2/N}$ we must use $\sqrt{\Sigma x^2/n-1}$ in order to calculate the SD. Here $n-1$, one less than the number in the sample, corresponds to the number of DF in a sample with a given mean (given the mean and $n-1$ of the items, one can calculate the n-th item). Using this rather than N as the numerator of the fraction will appreciably increase the value of the SD where N is small, and thus correct the underestimate of the population deviation which the simpler formula gives.

The SE formula remains the same

$$\frac{\text{SD}}{\text{square root of number of items in sample}}$$

but whereas when the population SD is used this SE can be used to divide the observed difference to give us the statistic z, when the sample SD is used the division of the observed difference by the SE will give us a different statistic t.[1] The probabilities associated with

$$z = \frac{\bar{x} - \mu}{\sigma/\sqrt{n}}$$

are quite different from those associated with

$$t = \frac{\bar{x} - \mu}{s/\sqrt{n}}$$

where the size of the sample is less than 30. With larger samples the probabilities associated with t approximate very closely to those associated with z, and the differences between the two can be ignored.

At the end of the book will be found a table of probabilities associated with t. Most textbooks of statistics devote considerable space to practice in the calculation of t and the evaluation of its probability:

---

[1] Often known as 'Student's t' after the pseudonym of its discoverer.

'*t*-testing' as it is called. For workers in the physical, social or psychological sciences, in which economic factors often force the use of samples smaller than 30, the differences between the values of z and the values of t are of great importance. In literary contexts the comparative cheapness of sampling means that the investigator will rarely be working with samples of less than 30 of the items he has selected for study. The fact that for small samples the distribution of t is considerably different from that of z is therefore of somewhat lesser importance.

Other statistics can be tested for significance by z-tests in the same way as proportions and means. The only difference is that each statistic has a different formula for calculating its characteristic SE. For the median, for instance, the formula is

$$SE = \frac{\sqrt{(\text{number of items in distribution})}}{2 \times \text{items in cell containing median/interval size}}.$$

However, the use of a z-test for the median can be very misleading in a skew distribution; whereas it is principally in skew distributions that the calculation of the median itself is of value. The same is true of quartiles. The use of z-tests for statistics other than the mean and proportion is of very limited utility.

Exercise 18.
You have a sample of 100 sentences with a mean length of 16.74 words and a SD of 8 words. (a) Test the hypothesis that this is drawn from a population with mean 15.14 words. (b) Test the hypothesis that this is drawn from the same population as another sample of the same size with mean 15.49 and SD of 6 words.
Solution.
(a) hypothesis rejected at 5% level ($t = 2$);
(b) hypothesis accepted ($t = 1.25$).

10

The Analysis of Variance

IN PREVIOUS chapters we have learnt how to assess the significance of various kinds of differences between sample statistics. We have learnt how to use the χ^2 test to evaluate the significance of differences between attribute data in any number of samples; we have used the z-test to compare differences between pairs of proportions, and the t-test to compare differences beween pairs of means. What we have not yet encountered is a method of testing for significance differences between means and proportions where more than two samples are involved. Such a method is given by the *analysis of variance.*

The analysis of variance is a collection of techniques for reducing the variation in a set of data into components associated with sources of variation. It thus permits the testing of the statistical significance between the means and proportions derived from several samples. It provides a method of testing the null hypothesis that the samples are independent samples drawn from populations having the same mean or, as it may be, proportion. Even if the null hypothesis is true, we know that there will be differences between the observed sample statistics due to sampling fluctuation. The analysis of variance assists us to decide whether the observed differences are greater than the fluctuation to be expected between random samples from similar populations.

Analysis of variance takes its start, as the name suggests, from an investigation of the *variance* to be observed in the data under study. The variance of a population, it will be remembered, was a measure of the variability of scores from the mean: it was defined as the mean

127

squared deviation, and was calculated by the formula

$$\sigma^2 = \frac{(X - \mu)^2}{N}$$

where N was the number of items in the distribution, X the individual score, μ the mean. When we wish to calculate the variance of a sample as an unbiased estimate of the variance of a population we have, as explained earlier, to use the denominator $n - 1$ instead of N.

If we have a number of samples we can compare the variances in the different samples; and we can compare these variances with the total variance to be found in the overall sample obtained by pooling the individual samples. The analysis of variance proceeds by separating or partitioning the total variance in the data into two parts: the variance *between* (or among) the samples, and the variance *within* the individual samples. Two estimates of the population variance are then calculated, one based on the variance between one sample and another, the other based on the variation of the values between each sample. If the null hypothesis is true, and the samples are drawn from populations having the same mean, then the difference between the two elements of the population variance should be no greater than the amount predictable from random sampling fluctuation. But if the variance between the groups is significantly greater than the variance within each group, then the samples cannot be regarded as coming from the same population.

The analysis of variance therefore consists of the following stages:

(1) The total variation of the data is ascertained. This involves calculating a grand mean for all the items in the pooled sample, and calculating the sum of the squares of the deviations of each of the squares from the grand mean.

(2) The variation between groups is calculated: this is the sum of the squares of the deviation of each sample mean from the grand mean.

(3) The variation within groups is calculated. This variation is the sum of the squares of the deviation of each score in each sample from the mean of its own sample; but it is most readily calculated by simply subtracting the between-groups variation from the total variation.

(4) The population variance is then estimated on the basis of the two types of variation just calculated. The variance estimates are called mean squares: that based on the variation between groups is called the mean square between groups, and that based on the variation within groups is called the mean square within groups.

When the analysis of variation is thus completed, the two estimates are compared with each other, and the ratio between them is evaluated for statistical significance. The ratio of the mean square between to the mean square within is called the F-ratio. If the null hypothesis is correct, the F-ratio can be expected to be near to unity; but if the groups genuinely differ from each other it will be greater than unity. Tables of significant values of the F-ratio enable us to determine whether the ratio observed is statistically significant.

Let us illustrate this in a simple example. Suppose we wish to study whether there is any significant difference between the number of words that different poets can pack into a heroic couplet; and to that end we select the three following five-couplet samples from three poets.

1. From Pope's *Essay on Criticism*

Avoid extremes; and shun the fault of such,
Who still are pleased too little or too much
At every trifle scorn to take offence,
That always shows great pride, or little sense;
Those heads, as stomachs, are not sure the best,
Which nauseate all, and nothing can digest.
Yet let not each gay turn thy rapture move;
For fools admire, but men of sense approve:
As things seem large which we through mists descry,
Dullness is ever apt to magnify.

2. From Johnson's *Vanity of Human Wishes*

Once more, Democritus, arise on earth
With chearful wisdom and instructive mirth,
See motley life in modern trappings dressed,
And feed with varied fools th'eternal jest:
Thou who couldst laugh where want enchained caprice,
Toil crushed conceit, and man was of a piece;
Where wealth, unloved without a mourner died,
And scarce a sycophant was fed by pride;
Where ne'er was known the form of mock debate,
Or seen a new-made mayor's unwieldy state;

3. From Goldsmith's *The Deserted Village*

'Tis yours to judge, how wide the limits stand
Between a splendid and a happy land.
Proud swells the tide with loads of freighted ore,
And shouting Folly hails them from her shore;
Hoards even beyond the miser's wish abound,
And rich men flock from all the world around.
Yet count our gains. This wealth is but a name
That leaves our useful products still the same.
Not so the loss. The man of wealth and pride
Takes up a space that many poor supplied;

We proceed as follows. First we calculate the grand mean, derived from the overall sample of 15 couplets. These couplets contain 240 words; dividing by 15 we get the grand mean, 16. We then record the deviation of each individual score from the grand mean, square each deviation and sum the squares. The calculations are shown in the first column of the table on the following page. The total SS is 32.

Next, we calculate the component of this total variability which is represented by the sum of squares between samples. We calculate the sum of the squares of the deviation of each sample mean from the grand mean. The squared deviation in each sample must be weighted by the number of couplets which it contains: that is to say, the squared deviation in each case must be multiplied by five, since each sample contains five couplets. This calculation is shown in the second column on the table on the following page. The between-groups sum of squares is 10.

We can then ascertain the within-groups variability simply by deducting the between-groups sum of squares from the total sum of squares, getting the result 22; however, it is also possible, as a check, to calculate the within-groups variability by squaring the deviation of each couplet from the mean from the poet who wrote it, and then summing these squares. This is done in the third column of the table on the following page, and gives again the result 22. It will be obvious that the samples have been chosen with a view to facilitating the calculation of the sums of the squared deviations.

The next stage is to make the two estimates of the population variance. The estimate is obtained by dividing the sums of squares by their appropriate DF. For the sum of squares between the DF are one less than the number of samples: as in the usual formula for an

Poet	Words	Calculation of Total sum of Squares		Calculation of Sum of Squares Between			Calculation of Sum of Squares Within	
		Deviation from grand mean	Squared deviation	Sample mean	Deviation from grand mean	Squared deviation (weighted)	Deviation from sample mean	Squared deviation
Pope	17	1	1	16	0	$0 \times 5 = 0$	1	1
	15	−1	1				−1	1
	16	0	0				0	0
	17	1	1				1	1
	15	−1	1				−1	1
	80		4					4
Johnson	12	−4	16	15	−1	$1 \times 5 = 5$	−3	9
	15	−1	1				0	0
	17	1	1				2	4
	15	−1	1				0	0
	16	0	0				1	1
	75		19					14
Goldsmith	16	0	0	17	1	$1 \times 5 = 5$	−1	1
	17	1	1				0	0
	16	0	0				−1	1
	18	2	4				1	1
	18	2	4				1	1
	85		9					4
Total	240		Total SS = 32			SS Between = 10		SS Within = 22
Gr Mean	16							

estimate of variance based on a sample it is

$$\frac{SS \text{ between}}{c - 1}$$

where c is the number of groups or samples. ('n' is here reserved for the number of items in the total distribution.) In this case, the estimate of population variance, that is to say the mean square between groups, is 10/2 or 5.

Calculating the DF for the population estimate based on the variability within groups is not quite so straightforward. We have three groups of five; the number of degrees of freedom in each group is four, and so the total for the three groups is 12. In general the formula for the estimate of variance based on within groups variation is

$$\frac{SS \text{ within}}{n - c}$$

where n is the number of items in the distribution and c the number of groups. In the present case, substitution in this formula yields a value of 22/12 or $1\frac{5}{6}$.

The F-ratio is given by the formula

$$F = \frac{\text{Mean square between}}{\text{Mean square within}} = \frac{5}{1\frac{5}{6}} = \frac{30}{11} = 2\frac{8}{11}.$$

We must finally discover whether the F-value is significant. If the null hypothesis is true, then the sample means will differ only by random fluctuation from the population mean, and the estimates based on the two different types of variability will be approximately the same, being estimates of the same common population variance. The ratio we have discovered is, however, considerably greater than unity. Whether it is significantly greater is to be found by consulting the F-table given in the Appendix.

The F-table gives the largest value of F that may occur due to sampling error at a given level of significance. Table A gives the values for the 5% level of significance, and table B gives the values for the 1% level of significance. To use the tables we need to know the DF for each of the mean squares. We first find the column corresponding to the degrees of freedom of the numerator of the F-ratio (mean square between) and then look down it to the row corresponding to the degrees of freedom of the denominator (mean square within). In the example worked above, we need to look to the column labelled 2 and the row labelled 12. We find that the value of F for the 5 per cent significance level is 3.88. Since this is considerably greater than our observed value of $2\frac{8}{11}$, we conclude that we cannot reject the null hypothesis, and that there is no evidence from these samples of significant differences between the number of words that particular poets pack into a heroic couplet.

The results of an analysis of variance (ANOVA) are commonly presented in a summary table. For the data analysed the table would read:

Source of variation	Sums of squares	DF	Mean squares	F	p
Between samples	10	2	5	2.73	>0.05
Within samples	22	12	1.83		

The calculation has been worked out above in a manner designed to bring out the logic of the analysis of variance. There exists a formula which achieves the same results with less cumbrous calculation. The total sum of squares may be calculated by the formula

$$SSt = \Sigma X^2 - \frac{(\Sigma X)^2}{N}$$

and the sum of squares between groups is calculated using the formula[1]

$$SSb = \frac{(\Sigma X_1)^2}{n_1} + \frac{(\Sigma X_2)^2}{n_2} + \frac{(\Sigma X_3)^2}{n_3} \cdots \frac{(\Sigma X_K)^2}{n_K} - \frac{(\Sigma X)^2}{N}.$$

The sum of squares within groups is obtained as before by subtraction of the sum of squares between from the total sum of squares, and F is calculated as before.

Let us illustrate the method of calculation on a different set of data. The following table gives, for samples from three works of Aristotle,

---

[1] Where X_K = value of an item in group K. and n_K = no. of items in group K.

percentage of sentences which have as their second word '*de*' ('but').

De Anima		De Caelo		Rhetoric	
Book	Percentage	Book	Percentage	Book	Percentage
I	55	I	32	I	35
II	50	II	34	II	21
III	39	III	37	III	34
		IV	43		

To carry out the analysis of variance, we can set out the sample proportions and their squares in columns in the table below, and then substitute in the formulae

De Anima		De Caelo		Rhetoric		Total
0.55	0.3025	0.32	0.1024	0.35	0.1225	
0.50	0.2500	0.34	0.1156	0.21	0.0441	
0.39	0.1521	0.37	0.1369	0.34	0.1156	
		0.43	0.1849			
1.44		1.46		0.90		3.80
	0.7046		0.5398		0.2822	1.5266

$$\text{Total sum of squares} = 1.5266 - \frac{14.44}{10} = 0.0826$$

$$\text{Sum of squares between} = \frac{(1.44)^2}{3} + \frac{(1.46)^2}{4} + \frac{(0.90)^2}{3} - \frac{14.44}{10}$$

$$= 0.6912 + 0.5329 + 0.2700 - 1.444$$

$$= 0.0501$$

Sum of squares within = $0.0826 - 0.0501 = 0.0325$
Mean square between = $0.0501/2 = 0.02505$; mean square within = $0.0325/7 = 0.00464$ F-ratio = $0.02505/0.00464 = 5.399$.
Reference to the F-table shows that this value is significant at the 5 but not at the 1% level.

It will have been obvious that the same result would have been reached had the percentages been treated as raw scores: the effect would have been to multiply all the entries in the first columns by 100 and in the second columns by 1000. This is, in practice, the more convenient method, avoiding a multiplicity of decimal points and zero's. However, the calculations were worked out for the actual proportions so as to illustrate that in computing the analysis of variance there is no real difference between the method for means and the method for proportions.

Where the data for the analysis of variance are simple, as in the first example worked in this chapter, there is little to choose between the two methods of calculation illustrated. But where the total mean and the sample means are not whole numbers, the calculation is much less cumbrous by the route just illustrated which makes use of raw scores rather than of deviations from the mean.

The analysis of variance explained in this chapter is the simplest method of such analysis: it is known as one-way analysis of variance. It is very commonly used in experimental circumstances in order to explore the effect of one independent variable, which the experimenter can manipulate, upon another variable which is surmised to be dependent upon it. Thus a doctor, wishing to compare the effects of various drugs, may compare variations in progress between groups of patients treated with the different drugs. In such an analysis of variance, the within-group variation would be that between individual patients each treated with the same drug: it would reveal the degree of sampling error to be expected. The between-group variation would be that between the means of the whole groups who had received different treatments. Because of the frequent use of analysis of variance in such contexts, the mean square between is sometimes called 'the treatments mean square' and the mean square within is called the 'error mean square', treatment and error being regarded as the two sources of variation. But the use of analysis of variance is not limited to such experimental contexts; and the name 'treatment' is clearly inappropriate in a literary context; it is mentioned here simply because it is frequently found in statistical textbooks.

In experimental contexts more elaborate methods of analysis of variance are practiced. For instance, the effect of a particular drug may be effected by whether it is used in conjunction with a different drug. A simple one-way analysis of variance exploring the effect of each drug in isolation would be inappropriate here. What is needed is a more elaborate technique known as two-way analysis of variance. This is beyond the scope of the present work. But one-way analysis of variance illustrates the method of partitioning a sum of squares into its parts which is the base of the methods used for the solution of problems more complicated than those illustrated here.

Exercise 19.

Perform an analysis of variance on the length of line, in words, of the following three Shakespearean passages and determine the F-ratio (count hyphenated words as single words).

1. It was the lark, the herald of the morn,
   No nightingale. Look, love, what envious streaks
   Do lace the severing clouds in yonder east.
   Night's candles are burnt out, and jocund day
   Stands tiptoe on the misty mountain tops.
   I must begone and live, or stay and die (*Romeo & Juliet*, III, v)

2. I know not, gentle men, what you intend,
   Who else must be let blood, who else is rank.
   If I myself, there is no hour so fit
   As Caesar's death hour; nor no instrument
   Of half that worth, as those your swords, made rich
   With the most noble blood of all this world. (*Julius Caesar*, III, 1)

3. Aye, in the catalogue ye go for men
   As hounds and greyhounds, mongrels, spaniels, curs
   Choughs, water-rugs and demi-wolves are clept
   All by the name of dogs. The valued file
   Distinguishes the swift, the slow, the subtle,
   The house-keeper, the hunter, everyone... (*Macbeth* III, 1)

Solution: F-ratio $= 75/26 = 2.88$; $p > 0.05$.

11

Theoretical Distributions and the Theory of Sampling

IN PREVIOUS chapters we introduced a number of tests of statistical significance. Each of these tests involved a decision at some point on the probability of the value of a certain statistic having been reached by chance. The probability, in each case, was evaluated by consulting a table. It is now time to explain the assumptions which lie behind the construction and use of such tables and the nature of the concept of probability on which statistical inference is based. The present chapter will briefly outline this concept and the assumptions implicit in its application, and consider how far these assumptions are verified in literary contexts.

The relevant notion of probability is best introduced by a consideration of games of chance. It was there, historically, that it took its origin. If we toss an unbiased coin we know that the chances of it coming up heads and it coming up tails are equal. These two cases, heads and tails, exhaust the possible outcomes: if the coin lands on its side we cannot count that as a toss. Since heads and tails are the only possible outcomes, and each is equally likely, we can say that the probability of each is one half, or 50%. If p is the probability of throwing a head, and q the probability of throwing a tail, we can say that p plus $q = 1$. In the mathematical treatment of probability, the total of the probabilities of the various possible outcomes of an event is by definition equal to one: where, as here, there are only two possible results, the probability of each is one minus the probability of the other: $q = 1 - p$ and $p = 1 - q$.

When an event has more than one possible outcome we may be particularly interested in one of these possibilities: we may, for

137

instance, bet on it. In such a case we might call such an outcome a 'success': the term is used very widely in statistical probability whether or not any bets have been laid or one outcome is intrinsically more desirable than another. If we bet, we may bet not just on single throws but on successions of throws: thus we might bet that there will be exactly two heads in three successive throws. In that case, several different possible outcomes would count as a success: HHT, HTH, THH. There are altogether eight possible outcomes of a series of three throws: HHH, HHT, HTH, HTT, THH, THT, TTH, TTT. Since each of these eight outcomes is equally likely, and three of the eight count as successes, the probability of success is $\frac{3}{8}$ or 37.5%.

A single event may, unlike the toss of a coin, have more than two possible outcomes. If we roll a die, there are six possible outcomes; if we draw a card blind from a pack there are 52 possible outcomes. We may bet on such an event in such a way that several different outcomes count as successes: for instance, we may bet that we will draw a spade. Since there are 13 different spades to be drawn, the probability of such a success is $\frac{13}{52}$ or 1 in 4, i.e. 25%.

In general we have

$$p = a/n$$

where p = the probability of success, n = the total number of equally possible outcomes, and a = the number of these that count as successes. Since the number of favourable outcomes can never be more than the number of possible outcomes, the value of p will always be between 0 and 1.

When we bet on the outcome of a series of independent events we can calculate the probability of success on the basis of the probabilities of the outcomes of the individual events. If we bet on a conjunction of independent outcomes, (a and b), then to obtain the probability of success we must multiply the probabilities of the outcomes together [$p(a) \times p(b)$]: the probability of getting heads in two successive throws is the probability of getting heads on the first throw multiplied by the possibility of getting heads on the second ($0.5 \times 0.5 = 0.25$). The probability of getting heads in three successive throws will be $(0.5)^3$ and in general the possibility of getting heads in n successive throws will be $(0.5)^n$.

What is the probability of drawing two successive spades from the pack? The answer to this depends on whether the first card is replaced after being drawn or not. If it is, then the second drawing is independent of the first, just as two successive tosses of a coin are independent of each other, and the probability is found by multiplying together the possibilities of the separate draws, i.e. = 0.25 × 0.25 or 0.0625. Performing independent trials analogous to this is called *sampling with replacement*. If, however, the card is not replaced, then the second draw takes place under conditions which have been altered by the first and is accordingly not independent of it. Suppose, for instance, that the first card drawn was a spade, which is not replaced. The second draw is then not from a pack with 52 cards and 13 spades, but from a pack with only 12 spades among 51 cards; the odds on getting a spade are no longer 1 in 4 but only 12 in 51 (0.235). To get the probability of drawing two spades, when sampling without replacement, we have to multiply together the two different probabilities thus:

$$p(a \text{ and } b) = p(a) \times p(b) = 13/52 \times 12/51 = 0.059.$$

When we are betting not on a conjunction of outcomes, but on a disjunction of outcomes, then the probabilities of the individual outcomes are not multiplied but added together. Thus, we may bet not that in two successive draws from a pack there will be both a spade and a club, but that in a single draw there will be either a spade or a club. The probability of a spade is $\frac{1}{4}$, and so is the possibility of a club; the possibility of a spade or a club is therefore $\frac{1}{2}$. The probability of a head or a tail on a single toss is $\frac{1}{2} + \frac{1}{2}$, i.e. one or certainty. This addition rule applies when, as in these cases, the individual outcomes are mutually exclusive (no card is both a spade and a club, no toss is both a head and a tail). Where the outcomes are not mutually exclusive then the addition rule must be modified. Suppose we consider the possibility that either a spade or an ace will be drawn. The probability of a spade is 1 in 4, and that of an ace 1 in 13. But we cannot add these probabilities together and say that the probability of a spade or an ace is $\frac{1}{4} + \frac{1}{13} = \frac{12}{52}$. For one of the possibilities is the drawing of the ace of spades, which is both a spade and an ace;

it is counted in determining the odds on spades, and in determining the odds on aces; so if we simply add the two probabilities together we will have counted this possibility twice. The probability of an ace or a spade is in fact $\frac{16}{52}$ (there are 13 spades, plus three aces from the other suits). In general when two alternatives are not mutually exclusive we must subtract from the total of their several probabilities the probability of an outcome combining both. The addition rule then becomes

$$p(a \text{ or } b \text{ or both}) = p(a) + p(b) - P(a \text{ and } b).$$

The calculation of the probabilities of different outcomes can be made perspicuous by being set out in a tree diagram such as that on the following page. This sets out, step by step, the possibilities of diamonds being thrown in four successive rolls of a four-sided die marked hearts, clubs, diamonds, spades. The penultimate column sets out all the possible outcomes; the final column gives their probabilities, calculated by the multiplication rule from the probabilities at each of the three steps. We can then apply the addition rule to work out the probabilities of any given success: for instance, the probability of rolling exactly two diamonds is the probability of DDN $(\frac{3}{64})$ plus that for DND $(\frac{3}{64})$ plus that for NDD $(\frac{3}{64}) = \frac{9}{64}$; the probability of rolling at least two diamonds is that plus the probability of DDD $(= \frac{9}{64} + \frac{1}{64} = \frac{10}{64})$.

The first step in the calculation of any of these probabilities is to work out the number of possible outcomes which count as successes. If a large number of events or trials are in question, it is not possible to do this by enumeration or display, as above. If we were considering a run of 10 rolls of our four-sided die, for instance, the final column in a table such as that above would consist of two to the tenth, or 1024, entries. Fortunately there exists a formula for determining how many successful outcomes there will be for any given number of events and given definition of success.

Suppose that n is the number of rolls of dice, and suppose that we decide that success is to consist in r heads. Then we need to know the number of combinations of r successes from n events. The mathematical symbol for this number of combinations is $\binom{n}{r}$ and the method of calculating it is given by the formula

$$\frac{n!}{r!(n-r)!}$$

where the sign '!' means 'factorial'. The factorial of a number is that number multiplied by the next smallest whole number, multiplied by the next smallest again, and so on until we multiply by one. Thus factorial 4 is $4 \times 3 \times 2 \times 1$, and factorial 5 is $5 \times 4 \times 3 \times 2 \times 1$.

If we apply the formula for combinations to our four-sided die to discover how many combinations of two diamonds there are in three rolls (substituting 2 for r and 3 for n)

$$\frac{3 \times 2 \times 1}{2 \times 1 \times 1} = 3,$$

First draw	Second draw	Third draw	Outcome probability
		D ($\frac{1}{4}$)	DDD $\frac{1}{64}$ = 1.6%
	D ($\frac{1}{4}$)	N ($\frac{3}{4}$)	DDN $\frac{3}{64}$ = 4.7%
Diamond (D) ($\frac{1}{4}$)		D ($\frac{1}{4}$)	DND $\frac{3}{64}$ = 4.7%
	N ($\frac{3}{4}$)	N ($\frac{3}{4}$)	DNN $\frac{9}{64}$ = 14.1%
		D ($\frac{1}{4}$)	NDD $\frac{3}{64}$ = 4.7%
	D ($\frac{1}{4}$)	N ($\frac{3}{4}$)	NDN $\frac{9}{64}$ = 14.1%
Not diamond (N) ($\frac{3}{4}$)		D ($\frac{1}{4}$)	NND $\frac{9}{64}$ = 14.1%
	N ($\frac{3}{4}$)	N ($\frac{3}{4}$)	NNN $\frac{27}{64}$ = 42.2%
			$\frac{64}{64}$ = 100.0%

as we discovered above by listing them. If we want to know how many combinations of five diamonds in ten rolls we apply the formula

$$\frac{10!}{5!(10-5)!} = \frac{10 \times 9 \times 8 \times 7 \times 6 \times 5!}{5! \times 5 \times 4 \times 3 \times 2 \times 1} = \frac{30240}{120} = 252.$$

In order to work out the probability of a successful outcome we have to take account not only of the number of combinations but also of the probability of each outcome at a single trial. In the case of our four-sided die, for instance, the probability of a diamond at each trial is only 1 in 4. We do so in accordance with the following formula:

$$\text{Probability} = \binom{n}{r} \times p^r \cdot q^{n-r}$$

where p is the probability of success on a single trial, and $q = 1 - p$. Thus the probability of exactly two diamonds in three rolls of a four-sided die is $3 \times 0.25^2 \times 0.75^{(3-2)} = 0.141$, or 14.1% as we worked out in our diagram. The probability of exactly two diamonds in six rolls is worked out by the same formula:

$$\frac{6!}{2! \times 4!} \times 0.25^2 \times 0.75^4 = 15 \times 0.0625 \times 0.316 = 0.297.$$

Where the probability of a success is equal to the probability of a failure at a single trial (as in tossing coins) the calculation is much simpler; for when p and q are both equal to 0.5, then $p^r \times q^{n-r}$ is always equal to $\frac{1}{2}^n$. (For $\frac{1}{2}^r \times \frac{1}{2}^{n-r} = \frac{1}{2}^{r+n-r}$.) Thus the probability of five heads in ten tosses of coins is

$$\binom{n}{r} \times \tfrac{1}{2}^n$$

where $n = 10$ and $r = 5$, that is to say $252/2^{10} = 252/1024$, or 0.246.

We could use our formula to calculate the probability of each possible successful outcome of ten tosses of a coin, from zero to ten heads. The probabilities can be set out in the following table:

Number of heads in ten tosses	Probability
0	0.001
1	0.010
2	0.044

3	0.117
4	0.205
5	0.246
6	0.205
7	0.117
8	0.044
9	0.010
10	0.001

we could represent these probabilities on a graph such as that on the following page.

The probabilities represented in the table above are known as binomial probabilities. A binomial probability is the probability of one out of two mutually exclusive and jointly exhaustive possible outcomes of an event. A graph such as that on the following page gives a binomial distribution: that is the distribution of r occurrences of successful outcomes of n events. The events in question must be independent of each other and each must have only two possible outcomes (either naturally so, as in the case of heads and tails, or because we have so grouped them, as in the case of diamonds vs non-diamonds). They may have two equally probable outcomes (as in heads vs tails) or have two outcomes of unequal probability (as in the case of diamonds on a four-sided die).

The tosses, throws, draws and deals of games of chance provide the paradigm of these binomial events (or Bernoulli trials, as they are called after the mathematician who first studied them); but there are countless other events which fulfil the definition of a binomial event and which can be studied by means of the binomial distribution. A birth, which may be the birth of a boy or the birth of a girl; a letter in a text, which may be either a vowel or a consonant; a word in a text, which may be or not be an 'and'; we can count each of these as an event with two possible outcomes, and count, if we will, the birth of a boy, the occurrence of a vowel, or the presence of an 'and' as a success.

The binomial distribution represented in the graph was calculated by applying the formula $\binom{n}{r} p^r q^{n-r}$ to a particular value of n, r and p. The expression $\binom{n}{r}$ is indeed known as the binomial coefficient because of this function of it. But it very quickly becomes tedious to work out binomial distributions in this way. Fortunately there exist

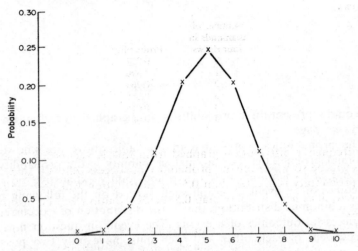

Binomial distribution of number of heads in ten coin-tosses.

tables which give binomial coefficients, and binomial probabilities, for different values of N and r. These are not given in this book because, even more fortunately, it is not necessary for our purposes to engage in the calculation of binomial probabilities once the concepts involved have been grasped.

Because of the way in which it is calculated any binomial distribution is completely described by the two parameters p and N. The mean of the distribution, i.e. the mean frequency of successes, is N $\times$ p, where N is the number of trials and p the probability; and the standard deviation of the distribution is $\sqrt{Npq}$ where $q = 1 - p$. For the number of heads in ten tosses of a coin, $N = 10$ and $p = 0.5$, so that the mean is 10×0.5, or 5, and the standard deviation is $10 \times 0.5 \times 0.5 = \sqrt{2.5} = 1.581$.

The binomial distribution of the tossing of coins was a symmetrical graph. This is because the probability of heads and tails was equal. If we graph the distribution of the number of diamonds thrown in five tosses of our four sided-die (or the number of diamonds drawn in five draws with replacement from a normal pack of cards) we find a different picture. The probabilities of zero to five diamonds are as follows:

Number of diamonds in four draws	Probability
0	0.237
1	0.396
2	0.264
3	0.088
4	0.015
5	0.001

The frequency distribution graphed from this is very skew in form. This will be so wherever the probability of success differs from 0.5: if the probability is greater than 0.5 the distribution will be negatively skewed; if, as here, it is less than 0.5 the distribution will be positively skewed.

However, whether the probability of success is equal to, or different from 0.5, given a sufficiently large *n*, the distributions will begin to approach the same form. This form, which is called the *normal* distribution, is of great importance in statistics: and to it we must now turn.

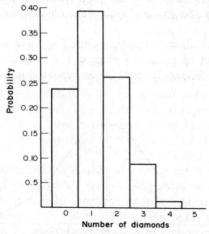

Binomial distribution for numbers of diamonds in five draws with replacement.

The normal distribution was developed in the early 18th century by the mathematician Abraham de Moivre in his studies of the probabilities of games of chance. He devised it as an approximation to facilitate the calculation of the distribution of chance events. The binomial distribution is a discrete distribution: that is to say it takes on only whole number values. There can be two or three tosses of a coin but there cannot be 2.87 tosses. In place of the jagged polygon or stepped histogram which represents the binomial distribution for a finite number of events, de Moivre developed a smooth continuous curve representing the form which the binomial distribution would take for an infinite number of events with equiprobable outcomes; and he showed both that this normal curve was much easier to calculate than the discrete distribution of the binomial, and that it provided a satisfactory approximation to the binomial distribution, even where the probabilities of the outcomes were not equal, provided that the number of events was relatively large.

The graph of a normal distribution is a bell shaped curve. It is symmetric and unimodal, so that the mean, the median, and the mode of the distribution all coincide. Its tails extend indefinitely to right and left, so that it is theoretically possible in a normal distribution to obtain values at any distance from the mean. The normal curve, as mathematicians say, approaches asymptotically to the zero axis.

The normal distribution has been found to have many applications outside the realm of games of chance. Carl Gauss (1777–1855) discovered that the discrepancies between repeated measures of astrono-

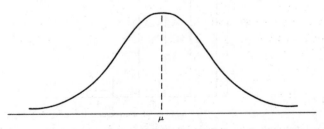

The normal distribution

mical phenomena were distributed in accordance with de Moivre's curve: accordingly he named it 'the normal curve of errors'. But it was soon found that many natural phenomena, and not just erroneous measurements of them, were distributed in accordance with the same curve: for instance, the heights of human males, the weights of animals of the same species and age, the results of psychological tests. Height and weight are themselves continuous variables which can take on any value that the accuracy of our measurement permits; hence the need for a continuous curve to represent their distribution. This is one reason why the normal distribution has a great importance in the physical and pyschological sciences: many of the data which are studied by those sciences are themselves values of continuous variables which are normally distributed. But that is not the only reason, as we shall soon see. If it were, then the normal distribution would be of little concern to students of literary statistics where few, if any, data are normally distributed.

The typical phenomena studied by the literary statistician are discrete phenomena—the occurrences of words, the number of letters in a word, the length of a verse in syllables. Distributions of such features as sentence length are highly skew and far from symmetrical, bearing little resemblance to the bell-shaped curve of the normal distribution. None the less the normal distribution is almost as important in the statistical study of literature as it is in the natural and social sciences. This is not because the *data* of the discipline are normally distributed, but because its *statistics* are.

One of the most important results of the mathematical theory of statistics is a theorem called the *central limit theorem*. This states that when samples are repeatedly drawn from a population the means of the samples will be normally distributed around the population mean. Most importantly, it states that this will be the case *whether or not the distribution of the data in the population is itself normal*.

Because the statistics—the means and proportions—of even non-normal distributions themselves tend to be normally distributed, the characteristics of the normal distribution can be used to study the sampling distribution of the statistics. The sampling distribution must be distinguished from both the sample distribution and the population distribution. A sample distribution is a distribution of the values of some variable ascertained by inspection of a sample: for instance, the distribution of sentence-lengths in a 100-sentence sample of *Paradise Lost*. The population distribution is the distribution of the values of the variable in the whole population which it is desired to study by means of the sample: for instance, the distribution of sentence-lengths in *Paradise Lost* as a whole. The sampling distribution is the distribution of the statistics produced by repeated sampling from the population: for instance, the distribution of the mean sentence-lengths of repeated one-hundred sentence samples of *Paradise Lost*.

Distributions may be actual or theoretical. When we record and classify and make a histogram of the lengths of sentences in a 100-word sample, we obtain the actual sample distribution. Commonly we do not know the actual population distribution of the phenomena we study: we make a hypothesis about the distribution, work out the properties of the distribution *a priori*, and then use samples to test whether the population conforms to this theoretical distribution. The sampling distribution that interests statisticians is the theoretical distribution of all possible values of a test statistic. Any sample statistic which we ascertain will have its place somewhere in this theoretical sampling distribution. All the tests of significance which we have studied in previous chapters have consisted in locating discovered statistics within the appropriate sampling distribution. For there is a probability associated with each location in a sampling distribution; and when we know where a test statistic is located in its sampling distribution we know how likely it is that it should have been produced as a result of random sampling from a population of a particular kind. The standard errors of statistics, which were of such importance in significance testing, are in fact the standard deviations of the relevant sampling distribution.

It is for this reason that it is important that sampling distributions are normal distributions. For in order to discover the probabilities associated with locations in sampling distributions, we have only to discover the probabilities associated with locations in the normal distribution.

There is a formula which enables one to derive the normal curve, that is to say, to determine the value on the *y* axis (the height of the curve) for any given value on the x axis (the base line for the curve).[1] All the expressions in the formula are constants except μ and σ, the mean and SD of the data in the distribution. Therefore, given these two parameters of the distribution, the mean and the SD, the appropriate normal curve and all its properties can be deduced. The normal distribution is in fact not a single distribution but a family of distributions. The distributions differ from each other by having different means and different SD, they belong to the same family because they are all derived from their mean and SD by the same formula.

In Chapter 4 we saw how a SD could be used to provide a measure of the location of each individual item in a distribution: we learnt to calculate the z-score of an item, that is to say, its distance from the mean measured in SD. If we convert every value in a distribution to a z-value, we obtain a SD which has a mean of 0 and a standard deviation of 1. If we apply this operation to a normal distribution, we obtain a standardized normal distribution, as illustrated below. This standard normal distribution can then be used to study the properties of normal distributions with any mean or any SD. In particular, it can be used to determine the probability of a particular location within the normal distribution, as will now be explained.

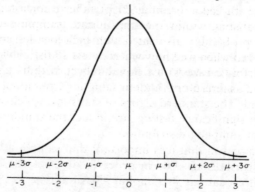

---

[1] The formula is

$$y = \frac{1}{\sigma \sqrt{(2\pi)}} e^{[-(X-\mu)^2/2\sigma^2]}.$$

Fortunately, the student has no need to commit it to memory. e is a constant, 2.71828.

Since the normal distribution is a continuous distribution, the probability of occurrence of any particular value in the distribution (e.g. the value 1) is zero; hence it is not the actual height of the curve (the y value) corresponding to a particular value which is relevant in determining probability. (What the height of the graph represents is the probability *density* at any given point along the x-axis.) What we must determine is the probability of obtaining a value within a particular interval of values, which corresponds to an *area* under a curve. The equation for the normal curve determines not only the points on the curve but the proportion of the total area under the curve which falls between any two points. Thus the area under any two points, z_1 and z_2, on the standard normal curve represents the frequency with which values between z_1 and z_2 occur, the total area, corresponding to the total frequency, being regarded as 1.

The student does not have to work out the areas under the normal curve: tables have been prepared, and are included in most statistics textbooks, giving the areas under the curve between any two points. From these tables the percentage of cases in the distribution included within any distance from the mean, measured in SD, can be ascertained. It is found that 68.26% of values in a normal distribution occur within a distance of 1 SD from the mean in either direction, and 95.46% of the cases within 2 SD in either direction. This can be seen in the diagram below: where it can also be seen that, for instance, the proportion of cases falling between +1 and +2 deviations from the mean—the hatched area—is 13.6%. This means that the probability of a case with this value occurring by chance within the distribution is 0.136 or 13.6%.

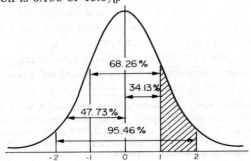

At the end of the book, a table of the area under the normal curve is given; the table of probability of SE unit deviations used in Chapter 7 was based on such a table. Because the sampling distribution for test statistics is normally distributed, we can use the normal curve to test the probability of the statistic's deviating by a certain amount from the value predicted by the null hypothesis. For the probability of a deviation of at least that size will correspond to the area under the normal curve which is more distant than it from the mean.[1] It is thus that the table of areas under the normal curve can be used to provide a table of the probability of SE unit deviations.

The use of the normal form in significance testing when the underlying distribution is not normal depends, of course, upon the central limit theorem. The central limit theorem states that test statistics tend to be normally distributed even when the samples which provide the statistics, and the populations from which they are drawn, are not normally distributed. But the sampling distribution is a close approximation to the normal distribution only when the number N, the sample size, is relatively large. If N is relatively small (say, less than 30) then the sampling distribution is not necessarily normal: if the underlying population is normal, and the population SD is known only by estimation from the sample SD, then the appropriate sampling distribution is the t-distribution which underlies the t-tests mentioned in Chapter 9. If the underlying distribution is not normal, then even the t-distribution will provide only an approximation to the sampling distribution. Since in literary studies however there is commonly no difficulty in assembling samples of larger than 50 for any significance test. For such cases the central limit theorem permits the use of tables of probabilities of SE unit deviations based on the normal curve even though it is known that the underlying distribution of stylistic phenomena is never normal.

In addition to its theoretical significance in literary statistics as underlying the use of significance tests, the normal distribution has a

---

[1] Because the normal curve has two tails, the area which gives the probability consists of the two extreme areas put together. Thus the normal curve area table shows that the area between the mean and a z of 2 is 0.4773 of the whole. The area then between $+2$ and -2 is 0.9546, and the probability of a deviation greater than 2 SE units in either direction is $1 - 0.9546$ or 0.0464.

further use as providing an approximation to the underlying non-normal distributions for sufficiently large samples. It has already been mentioned that the normal distribution started life as an approximation to the binomial distribution of events occurring in games of chance.

The larger the number of items in the sample, the closer the approximation between the binomial and the normal distribution. Indeed, if *n* were increased to infinity the two curves would coincide. But even with comparatively low values of *n* the approximation is good where the probability of success is close to 0.5.

We can illustrate this in an example. In written English letters from the first half of the alphabet (A–M) occur with approximately the same frequency as those from the second half of the alphabet (N–Z). For any given letter the probability of its being from the first half of the alphabet is approximately 0.5. Suppose that we divide a text into 10-letter segments and put a 0 for every letter from the first half of the alphabet, and a 1 for every letter in the second half of the alphabet. What is the probability that any given segment will provide exactly seven zeros?

We can calculate this using the binomial formula. The probability is

$$\frac{10!}{7! \times (10 - 7)!} \times 0.5^7 \times 0.5^3 = 120 \times 0.5^{10} = 0.117.$$

But instead of using the binomial formula we could calculate by means of the normal curve. We could treat the binomial distribution as a normal distribution with the same mean and SD. The mean of a binomial distribution is given by the formula

$$\mu = Np$$

that is to say, the mean is the number of trials multiplied by the probability of success, in our case 10×0.5 i.e. 5. The standard deviation of a binomial distribution is given by the formula

$$\sigma = \sqrt{Np(1 - p)}Np(1 - p)$$

that is to say, it is the root of the product of the number of trials and the probability of the two outcomes; in this case

$$\sqrt{10 \times 0.5 \times 0.5} = \sqrt{2.5} = 1.581.$$

In order, then, to use the normal approximation to our binomial distribution, we treat the binomial distribution as a normal distribution with mean of five and SD of 0.581. We want to find the probability of a score of seven; but since the normal distribution is a continuous distribution, we have to treat this, in accordance with our usual method for treating discrete values as continuous ones, as a score falling between 6.5 and 7.5. We must next turn the scores into z-scores by dividing by the SD as a unit of measurement. A score of 6.5 is a z-score of $6.5 - 5/1.581 = 0.949$; a score of 7.5 is a z-score of $7.5 - 5/1.581 = 1.581$. The area under the normal curve between a z-score of 0.949 and one of 1.581 is 0.114.

The probability thus calculated is quite close to the exact binomial probability already calculated (0.117). The approximation would be even closer if we took 20-letter segments of text and wished to calculate the probability of 14 zeros. Here we have a binomial distribution with mean of 10 and SD of 2.236; the exact binomial probability is 0.03696. The corresponding probability of a z-score between 1.565 ($= 3.5/2.236$) and 2.01 ($= 4.5/2.236$) is 0.03657. This illustrates how closely the normal curve approximates the binomial distribution when n is as high as 20 and p is close to 0.5.

The approximation is not as close when p is much greater or less. Suppose that we are interested in the occurrence of a frequent word, such as 'and', and suppose that we have reason to believe that it occurs with a frequency of one in 20 words, or 0.05. If we divide the text again into 20-word segments, we can calculate the probability of a segment containing 'and' exactly twice by the binomial formula and obtain the result: 0.189. For the corresponding normal distribution, with mean 1 and SD 0.975, the probability of a score between 1.5 and 2.5, that is of a z score between 0.513 and 1.538 is $0.304 - 0.062 = 0.242$; this greatly overestimates the exact binomial probability. But with 50-word samples the approximation would not be too bad: by the exact binomial method the probability of such a segment containing 'and' exactly five times is 0.066 by the normal approximation it is 0.071. With 100-word samples the probability of exactly five 'ands' occurring is, by the exact binomial method, 0.180; by the normal approximation, 0.183.

In the majority of cases it is far easier, if the calculation is to be done by hand, to work out the probability by looking up normal curve tables than it is to work out the exact binomial probability. This is especially so if we wish to work out the probability of obtaining a result *greater than* a particular number of successes. To work out the binomial probability of obtaining seven *or more* 'ands' in a 20-word segment one would have to work out separately the probability of seven 'ands', eight 'ands', nine 'ands', and so on through to 20 'ands', and then add these probabilities together. The normal approximation, on the other hand, can be worked out by a single consultation of the normal area tables. The greater the value of *n* the greater is the amount of labour saved by using the normal approximation; this is particularly gratifying since the greater the value of *n* the more accurate the normal approximation. However, when the calculations are done by calculator or computer, there is little to choose between the labour involved.

We have seen that when the probability of success is as low as 0.05 the normal approximation to the binomial is not accurate. Indeed, with low probabilities, the binomial distribution itself is no longer the appropriate one to describe the frequencies of the events involved. Its place is taken by a different discrete distribution, the Poisson distribution. The formula for the terms of a Poisson distribution may be written thus:

$$pr(r) = \frac{(np)^r}{r! \times e^{np}}$$

where *e* is a constant equal to 2.71828. The formula gives the probability of *r* events where *n* is the number of trials and *p* the probability of success at each trial. Once again, the student need not worry about applying this fearsome formula himself, since tables have been prepared for the Poisson distribution, tables which are entered through the value of *np* and which give the probabilities for values of 0, 1, 2, etc. of *r*.

Just as the binomial distribution approximates to the normal distribution for high values of *n*, so the Poisson distribution approximates to the binomial distribution as n increases; the approximation is close for values of $np = 5$ or more. For our example of the occurrences of 'and' in 100-word segments, the value of np is in fact 5, that is to say 100×0.05. The probability of such a segment containing exactly five 'ands', calculated according to the Poisson distribution, is 0.1755. The binomial probability, already calculated, was 0.1800.

A difference between the binomial and the Poisson distribution is this. The variance of the binomial distribution is npq. Since q can never be more than 1, the variance of the binomial can never be greater than the mean, which is np. The variance of a Poisson distribution, however, is not npq but np: hence it is equal to the mean which is also np. This can be used as an indication of the distribution of a set of rare events: if the variance is less than the mean, they most likely fit a binomial distribution; if the variance is equal to the mean, they most likely fit a Poisson. (If the variance is greater than the mean, there is another simple distribution which might be appropriate, the negative binomial, which will not be considered here.)

The way to test which theoretical distribution is fitted by a particular set of empirical data is to employ the χ^2 test introduced in Chapter 7 above. There the test was used to measure the differences between values observed in subsamples and values predicted from the constitution of an overall sample comprising all the subsamples: the test can be used also to evaluate the differences between values observed in samples and values predicted from a theoretical distribution such as the normal, binomial or Poisson. This is called a 'goodness-of-fit' χ^2 test.

Suppose that we have 100 20-word segments from a text in which, as a whole, 'and' occurs with a frequency of 0.05, or once in every 20 words. We examine each segment and observe the following distribution.

No of 'ands' in segment	Number such segments
0	40
1	35
2	17
3	7
4	1
5 or more	0

The theoretical distribution, calculated by the binomial probability formula, is as follows:

No of 'ands' in segment	Number such segments
0	35.8
1	37.8
2	18.9
3	5.9
4	1.3
5 or more	0.3

We can calculate χ^2 by the usual formula

$$\chi^2 = \sum \frac{(O - E)^2}{E}.$$

Expected value	Observed value	$O - E$	$(O - E)^2/E$
35.8	40	4.2	0.44
37.8	35	-2.8	0.22
18.9	17	-1.9	0.21
7.5	8	-0.5	0.03
			0.90

(The last three classes have been grouped together in accordance with the rule that no cell should contain less than five items). The value of χ^2 (0.90 for three degrees of freedom) is not statistically significant and so we conclude that our data provide a good fit to the binomial distribution.

The purpose of fitting data to theoretical distributions is to facilitate the description of the distribution. The form of a normal distribution is known once the mean and SD are known; that of a binomial distribution once the number of trials and probabilities at each trial are known; that of a Poisson distribution once the value of np is known. On the other hand, to describe the form of a skew distribution such as a sentence-length distribution, a large number of statistics are necessary: medians, first and third quartiles, ninth deciles and the like.

However, there are ways in which non-normally-distributed data can be transformed so that the normal distribution can be used to describe them. For instance, there are distributions in which the values are not normally distributed but in which it is found that the *logarithms* of the values are normally distributed. In such a case we have what is called a log-normal distribution.

It will be recalled that the distributions of word-length in letters investigated in the 19th century by T. C. Mendenhall were markedly skew. C. B. Williams in the present century has shown that these distributions, and many sentence-length distributions also, are approximately log-normal. The table below gives the distribution of word-lengths in letters in *Oliver Twist* estimated by Williams on the basis of Mendenhall's graphs: it also gives the logarithms of each of the letter-length values. The two frequency polygons show the distributions of the actual lengths and of the logarithms of the lengths. It will be seen that the skew distribution of the first graph is transformed into a visibly much more symmetrical form in the second.

Number of letters per word	Logarithm of this number	Number of such words per 1000 words
1	0	40
2	0.30	155
3	0.48	247
4	0.60	192
5	0.70	112
6	0.78	90
7	0.85	60
8	0.90	38
9	0.95	30
10	1	15
11	1.04	8
12	1.08	5
Over 12		8

Frequency of usage, per 1000 words, of words of different length in a 10,000-word sample of *Oliver Twist*; from Williams (*Style & Vocabulary*, 38) after Mendenhall.

There is no need, in order to represent graphically a log-normal distribution, actually to go through the step of ascertaining the logarithm of each value and then plotting it accordingly on ordinary graph paper. Graph paper is obtainable on which the x-axis is on a logarithmic scale, thus

<u>1   2   3   4   5   6   7   8 9 10   20</u>.

If the values are plotted onto such paper, then the logarithmic trans-

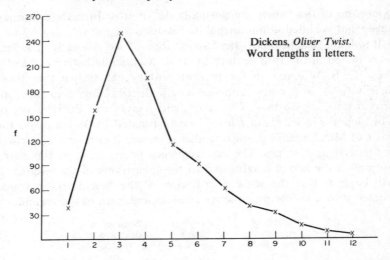

Dickens, *Oliver Twist.*
Word lengths in letters.

form of the histogram is automatic. There exists also a form of graph paper on which a logarithmic scale appears along the *x*-axis while the *y*-axis plots the cumulative total of values on a scale calibrated in accordance with the probabilities corresponding to the areas under

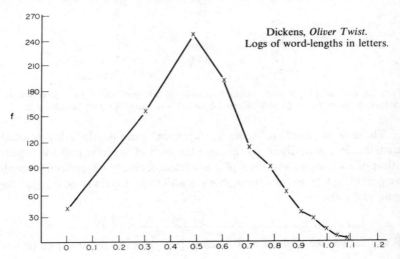

Dickens, *Oliver Twist.*
Logs of word-lengths in letters.

the normal curve. Such a graph, known as a log-probability graph, offers an easy test of log-normality, since a log-normal distribution will be indicated by a straight line.

Exercise 20.

Use the normal curve area table to find the probability of a z-score equal or greater than the following values. (a) 1.0, (b) 1.75, (c) 2.95. Check your solutions against the table on p. 103.

12

The Practice of Literary Sampling

A READER who has persevered thus far will be anxious to begin to apply the techniques described in this book to literary material in the field of his own interest. This final chapter deals with the selection, preparation and presentation of textual material in stylometric studies of literature.

The statistical theory on which significance testing is based concerns the relationship between random samples and the populations from which they are drawn. The fragments of text which have been used in this book to illustrate statistical techniques are not at all randomly chosen samples. They were carefully selected as self-contained pieces displaying particular characteristics and exhibiting data whose statistics could be calculated with ease. A random choice of literary material in English would have produced a very much less manageable set of texts.

If a fragment of text is to be chosen, not to illustrate a statistical technique, but in order to confirm or refute a statistical hypothesis, then it must be chosen in a very different way. A sample can be taken as representative of a population only if it is, in a technical sense, a random sample: that is to say, a sample chosen in such a way that every possible sample has an equal chance of being selected. The theorems stated in the second part of this book about the relationship betwen samples and population are valid only for the relationship between *random* samples and the whole population. Consequently we must consider the ways in which samples can be chosen in order to produce random samples.

In literary contexts there are three principal kinds of relationship between sample and population that interest us. In the first, and

161

simplest case, we wish to make generalizations about a large existing population without being put to the trouble of investigating it throughout; for instance, we wish to estimate stylistic parameters of *War and Peace* on the basis of a comparatively small fragment of the work. In the second case we study both the parameters of an existing whole, and the statistics of particular parts, and our question is whether the parts in question exhibit the features of samples chosen at random from the whole. We might, for instance, study the speeches of female characters in a play of Shakespeare to see whether the relationship between the quantitative features of these speeches and quantitative features of the whole play were such as were to be expected between randomly chosen sample and a whole population. In the third case, the population is not an actual one, but a hypothetical one, and we compare two existing texts to see whether the statistics they exhibit are those to be expected from two different samples drawn from the same population. This is one typical situation in authorship studies: we have an undisputed corpus of a particular author, and a controverted text, and we look to see whether the statistics of the corpus and the text are such as could be expected in two samples from a particular population. Here the population is not the *actual* corpus of the author's work—indeed, given the controversy, the bounds of that population are precisely what is in question; the population is rather a hypothetical population, the totality of what the author *could* have written consistently with the stylistic habits exhibited by the undisputed corpus. In these three cases the notions of 'population', 'sample' and 'choice' are used in rather different ways.

The first case is the clearest and provides the paradigm for the other cases. Before choosing a random sample as representative of an actual population the first step is to define clearly the limits of the population. If the population to be studied is, say, the pronouns in *Antony and Cleopatra*, it is important to decide and to record what edition of the play is being used, whether stage directions are included or excluded, whether possessive adjectives such as 'my' count or do not count as pronouns, and so forth. Only if such decisions are explicitly taken and clearly recorded will other scholars be able to check and utilize the results of one's study.

Once the population is defined, the sample must be chosen. It is not difficult to choose a random sample from a text, but it is not quite as easy as it may sound. It will not do, say, to pick the first 100 lines or sentences and hope that they are typical of the whole. Authors very often write quite differently at the start of a work: the gospels of Luke and John provide obvious examples of this. Nor will it really do to open a book at random, as in *sortes Virgilianae*, and start transcribing from the top of a page; books very often open at familiar and well-thumbed passages which may be loved precisely because they are uncharacteristic. The simplest method of producing really random samples from a text is to number the items of the text (words, lines or sentences) and then choose by a random process a number from which to start one's sample. It is not necessary to toss coins or roll dice to determine the number of the starting point: mathematicians have prepared tables of random numbers, one of which is included at the back of the book, which simulate the products of such randomizing devices.

Thus, suppose one wished to choose a 100-line sample from *Samson Agonistes* on which to base generalizations about the poem as a whole. The poem contains 1758 lines. One way to choose a random sample would be to take a table of random numbers and start sampling at the line indicated by the first four digits in the table which fell within the limits 1–1758. Then if the table begins

57494   72484   22676   44311   15356   05348...

one would start sampling at line 1115, whereas if the table began

92655   62097   81276   06318   81607   00565...

one would start sampling at line 978, since 0978 is the first set of four digits between 0001 and 1758. Sampling in this way is sometimes called *block sampling*. Another way of choosing a 100-line sample would be to choose 10 different random numbers and use them as the starting points of 10-line fragments, or to choose 100 different random numbers and make the sample up of these 100 disconnected lines. This method is sometimes called *spread sampling*. Which of these methods is preferable will depend, among other things, on what the phenomenon under study is: the sample of 100 disconnected lines will obviously be useless for a study of sentence-length, while it may be the most accurate indicator of word-frequency.

A disadvantage of block sampling is that the block chosen may be unrepresentative in a number of ways. For instance, if the number chosen to start sampling *Samson Agonistes* were anywhere between 1 and 14, the entire sample would consist of a speech of Samson himself. If Milton varies his style from character to character, it would be quite unsafe to generalise from this sample to properties of the poem as a whole. One way to avoid this difficulty is to adopt *stratified* random sampling.

Stratified random sampling is the kind of sampling which is familiar to us from the work of opinion pollsters. When a pollster wishes to predict the result of an election, he does not simply poll the first 100 people he meets in the street, or pick out a 100 voters from the electoral roll with a table of random numbers. He first divides the population into sex, age, geographical, social and economic groups, and then constructs a sample within which these different sections will be represented in the same proportion as they are in the population as a whole. Within the sections he will choose the members of the sample by random methods; his final sample will be a stratified random sample, not a simple random sample.

To use stratified random sampling in a study of a play by Shakespeare one might decide that one's sample was to contain so many lines of verse and so many of prose; so many lines spoken by male characters and so many by female characters; so many lines from each act of the play, and so on. Having devised one's stratification, one would use a table of random numbers to choose the items for inclusion in each category. Which categories one chose in the stratification would depend on what features one thought might be relevant to the stylistic phenomenon one was studying, just as the pollster stratifies the population in accordance with the features which experience shows to be relevant to voting behaviour.

So much for the choice of samples to represent a population. In the other two cases there is no genuine act of random choice. When we compare a whole with selected parts, we are not extrapolating to the whole from parts chosen to be random representatives: we study the population direct. And we choose the parts on the basis of our interests, not by any random process. In an authorship study, we are certainly not asking whether the author chose the words of a disputed text by using a table of random numbers to select from his entire vocabulary. But in each of these two cases, we are comparing statistics of fragments which we know were *not* chosen as random samples to see how far they differ from statistics to be expected *if they had been.* If they do not significantly differ, then we know that we do not need to look for particular kinds of explanation of the statistics we find. If the statistics of the speeches of female characters are what would be expected in a random sample from the play of the appropriate length, then we do not need to postulate a decision of the author's to vary his style in accordance with the sex of his characters; if there are no differences between the statistics of an undisputed corpus and those of a controversial work which would rule out the hypothesis that these were two random samples from a homogeneous population, we have so far no grounds for questioning the authenticity of the disputed text.

The samples on which statistical studies of literature are based are rarely completely random samples. A. Q. Morton has written, *a propos* of block sampling:

> Block samples are not usually random samples; they would only be random samples if every observation after the first one which starts the sample were independent of those before and after it. In literary texts this is not often the case. The reason for this is that language is not random in its fine detail, and can only be treated as random when the sample is large enough to eliminate periodic effects. That language is not random is simply illustrated. On the average, one word in twenty in all Greek writing is a repetition of the conjunction *kai*. If language were random in structure then once in every twenty times twenty words, i.e. once in every four hundred words, two occurrences of *kai* would come together. In fact no such pair of successive occurrences has yet been found in a Greek text. (*Literary Detection*, p. 76.)

This might seem to cut the ground from beneath the use of sampling techniques in literary studies; but in fact it does not do so. We need to look more closely at the concept of *independence*. It is true that in block sampling the choice of the second item to be observed is not independent of the choice of the first item, made on the basis of a random number; indeed the choice of every item depends on the choice of the first. However, this does not prevent the sample from being a random one in the sense that it was chosen by a method which gave every possible sample an equal chance of being selected.

There is no doubt, however, that the author's choice of one word in a text affects his choice of the next word. Does not this mean that the succession of words which constitutes a literary text is unsuitable material for statistical techniques based on successions of independent coin tosses or draws of cards? Not necessarily. The use of statistics to study the distribution of successive or concomitant events does not presuppose that these events are causally independent of each other. If determinism is true, then no two events are causally independent; but the use of statistical methods does not presuppose disbelief in determinism. The sense in which events must be independent if certain statistical laws are to apply to them is two-fold. In the first place, two events are independent only if there is no possibility of a single outcome counting as each of them. (It is in this sense that the drawing of a red card from the pack is not an event independent of the drawing of a court card from the pack.) In the second place, two events are independent only if the *a priori* possibilities of each are unaffected by the occurrence of the other. (It is in this sense that when cards are drawn without replacement the probabilities of later draws are affected by the outcome of earlier draws: the probability of drawing a red card on the second draw is reduced if a red card has been drawn on the first.) There is a third sense in which two events may fail to be independent: *experience may show* that the probability of one is affected by the occurrence of the other. But this kind of independence is not something that rules out the application of statistical methods; on the contrary it is something which statistical methods enable us to discover and predict.

It may be objected that the improbability of the occurrence of 'and' or its equivalent in other languages is not just an experimental discovery: it is something known *a priori* by anyone who understands the role of conjunctions in language. This is indeed true, and it is something which would have to be taken into account by anyone who wished to use statistical methods to study the serial features of language. But the methods illustrated in this book take no account of language's serial nature. A sentence-length distribution, or a word frequency list, does not preserve the order of words or sentences; it is as if words or sentences were cut out of a text and then shuffled together in a bag. Consequently, the only influence of the occurrence of an 'and' which we need to take account of is not its influence on the next word in a text, but its influence on a word occurring anywhere at all in the text; which is negligible.

The fact that the statistical methods used in this book (and in the great majority of published works of stylometry) do not take account of the serial nature of language in no way invalidates their use. What it does mean is that a great deal more information about stylistic features remains to be studied even after these techniques have been exploited to the full. There are sophisticated statistical techniques for the study of this further information, but they involve complicated mathematics well beyond the scope of the readers and the writer of this book.

The illustrations and the elementary exercises in this book have been based on counts carried out by hand and calculations performed with pencil and paper. Nowadays any serious student of literary stylometry will make use of electronic aids both in the assembling of data about his text and in the calculation of the statistics which interest him. There already exist a number of published concordances and word counts of the words of major authors, produced by computer program; there exists a computer language (SNOBOL) specially designed to facilitate computer handling of literary texts. It is not difficult for even the most innumerate to learn how to use the computer well enough to write programs to produce sentence-length distributions or the like. Learning to program a computer is, for instance, very much easier than learning to drive a car. But students of literature can use computers to produce word-counts and concordances without even learning any programming-language, since there exist packages of pre-written programs for this purpose. A package much used in Great Britain is the COCOA concordance and word-count package; this is now being superseded by OCP, the Oxford Concordance Package. There is an excellent guide to computer concordancing, and other uses of computers in literary studies, in Susan Hockey's book *A Guide to Computer Applications in the Humanities* (Duckworth 1980).

Just as there are packages to take the pain out of concordance-making, there are also packages to save the student the tedium of statistical calculation. All the techniques described in this book feature in an internationally used package called SPSS (Statistical Package for Social Scientists); the present book could be used in conjunction with the SPSS handbook to show how the programs in the package can be exploited for use in the literary and humanistic disciplines. The development of desk and pocket calculators has also meant that for small sample work it is unnecessary to go as far as a computer: most of the techniques in this book, for instance, can be performed on the TI 58/59 Texas Instruments calculator with the applied statistical module.

Between calculators and computers there are microcomputers some of which, such as the Intertec 'Superbrain' and the North Star 'Horizon', have the capacity to run concordance and word count programs. However, the price of these microcomputers still puts them out of reach of the private user. Moreover, SNOBOL, the most convenient programming language for literary and linguistic students, is not yet implemented on any microcomputer. However, given the rapid pace of development in this area, we can hope that the day is not long distant when the student of literature can perform the computation of style on his own microcomputer in his own study.

TABLE 1. Percentage points of the χ^2
distribution*

d.f.	P = 0.050	0.025	0.010	0.001
1	3.841	5.024	6.635	10.828
2	5.991	7.378	9.210	13.816
3	7.815	9.348	11.345	16.266
4	9.488	11.143	13.277	18.467
5	11.071	21.833	15.086	20.515
6	12.592	14.449	16.812	22.458
7	14.067	16.013	18.475	24.322
8	15.507	17.535	20.090	26.125
9	16.919	19.023	21.666	27.877
10	18.307	20.483	23.209	29.588
11	19.675	21.920	24.725	31.264
12	21.026	23.337	26.217	32.909
13	22.362	24.736	27.688	34.528
14	23.685	26.119	29.141	36.123
15	24.996	27.488	30.578	37.697
16	26.296	28.845	32.000	39.252
17	27.587	30.191	33.409	40.790
18	28.869	31.526	34.805	42.312
19	30.144	32.852	36.191	43.820
20	31.410	34.170	37.566	45.315
21	32.671	35.479	38.932	46.797
22	33.924	36.781	40.289	48.268
23	35.173	38.076	41.638	49.728
24	36.415	39.364	42.980	51.179
25	37.653	40.647	44.314	52.620
26	38.885	41.923	45.642	54.052
27	40.113	43.194	46.963	55.476
28	41.337	44.461	48.278	56.892
29	42.557	45.722	49.588	58.302
30	43.773	46.979	50.892	59.703
40	55.759	59.342	63.691	73.402
50	67.505	71.420	76.154	86.661
60	79.082	83.298	88.379	99.607
80	101.879	106.629	112.329	124.839
100	124.342	129.561	135.807	149.449

* Abridged from Table 8 of E. S.
Pearson and H. O. Hartley, *Biometrika Tables for Statisticians* vol. 1,
Cambridge University Press, 1954, by
permission of Professor E. S. Pearson
on behalf of the Biometrika Trustees.

TABLE 2. The variance ratio (F) 5% points* $(P = 0.05)$

n_1/n_2	1	2	3	4	5	6	8	12	24	∞
3	10.13	9.55	9.28	9.12	9.01	8.94	8.84	8.74	8.64	8.53
4	7.71	6.94	6.59	6.39	6.26	6.16	6.04	5.91	5.77	5.63
5	6.61	5.79	5.41	5.19	5.05	4.95	4.82	4.68	4.53	4.36
6	5.99	5.14	4.76	4.53	4.39	4.28	4.15	4.00	3.84	3.67
7	5.59	4.74	4.35	4.12	3.97	3.87	3.73	3.57	3.41	3.23
8	5.32	4.46	4.07	3.84	3.69	3.58	3.44	3.28	3.12	2.93
9	5.12	4.26	3.86	3.63	3.48	3.37	3.23	3.07	2.90	2.71
10	4.96	4.10	3.71	3.48	3.33	3.22	3.07	2.91	2.74	2.54
11	4.84	3.98	3.59	3.36	3.20	3.09	2.95	2.79	2.61	2.40
12	4.75	3.88	3.49	3.26	3.11	3.00	2.85	2.69	2.50	2.30
13	4.67	3.80	3.41	3.18	3.02	2.92	2.77	2.60	2.42	2.21
14	4.60	3.74	3.34	3.11	2.96	2.85	2.70	2.53	2.35	2.13
15	4.54	3.68	3.29	3.06	2.90	2.79	2.64	2.48	2.29	2.07
16	4.49	3.63	3.24	3.01	2.85	2.74	2.59	2.42	2.24	2.01
17	4.45	3.59	3.20	2.96	2.81	2.70	2.55	2.38	2.19	1.96
18	4.41	3.55	3.16	2.93	2.77	2.66	2.51	2.34	2.15	1.92
19	4.38	3.52	3.13	2.90	2.74	2.63	2.48	2.31	2.11	1.88
20	4.35	3.49	3.10	2.87	2.71	2.60	2.45	2.28	2.08	1.84
21	4.32	3.47	3.07	2.84	2.68	2.57	2.42	2.25	2.05	1.81
22	4.30	3.44	3.05	2.82	2.66	2.55	2.40	2.23	2.03	1.78
23	4.28	3.42	3.03	2.80	2.64	2.53	2.38	2.20	2.00	1.76
24	4.26	3.40	3.01	2.78	2.62	2.51	2.36	2.18	1.98	1.73
25	4.24	3.38	2.99	2.76	2.60	2.49	2.34	2.16	1.96	1.71
26	4.22	3.37	2.98	2.74	2.59	2.47	2.32	2.15	1.95	1.69
27	4.21	3.35	2.96	2.73	2.57	2.46	2.30	2.13	1.93	1.67
28	4.20	3.34	2.95	2.71	2.56	2.44	2.29	2.12	1.91	1.65
29	4.18	3.33	2.93	2.70	2.54	2.43	2.28	2.10	1.90	1.64
30	4.17	3.32	2.92	2.69	2.53	2.42	2.27	2.09	1.89	1.62
40	4.08	3.23	2.84	2.61	2.45	2.34	2.18	2.00	1.79	1.51
60	4.00	3.15	2.76	2.52	2.37	2.25	2.10	1.92	1.70	1.39
120	3.92	3.07	2.68	2.45	2.29	2.17	2.02	1.83	1.61	1.25
∞	3.84	2.99	2.60	2.37	2.21	2.10	1.94	1.75	1.52	1.00

* Abridged from Table 5 of Fisher and Yates, *Statistical Tables for Biological, Agricultural and Medical Research*, Oliver & Boyd, Edinburgh, by permission of the publishers.

TABLE 3. *Area under the normal curve, from O to Z.*

z	0	1	2	3	4	5	6	7	8	9
0.0	0.0000	0040	0080	0120	0160	0199	0239	0279	0319	0359
0.1	0.0398	0438	0478	0517	0557	0596	0636	0675	0714	0753
0.2	0.0793	0832	0871	0910	0948	0987	1026	1064	1103	1141
0.3	0.1179	1217	1255	1293	1331	1368	1406	1443	1480	1517
0.4	0.1554	1591	1628	1664	1700	1736	1772	1808	1844	1879
0.5	0.1915	1950	1985	2019	2054	2088	2123	2157	2190	2224
0.6	0.2257	2291	2324	2357	2389	2422	2454	2486	2517	2549
0.7	0.2580	2611	2642	2673						
					2704	2734	2764	2794	2823	2852
0.8	0.2881	2910	2939	2967	2995	3023				
							3051	3078	3106	3133
0.9	0.3159	3186	3212	3238	3264	3289				
							3315	3340	3365	3389
1.0	0.3413	3438	3461	3485	3508					
						3531	3554	3577	3599	3621
1.1	0.3643	3665	3686	3708						
					3729	3749	3770	3790	3810	3830
1.2	0.3849	3869	3888	3907	3925					
						3944	3962	3980	3997	4015
1.3	0.4032	4049	4065	4082	4099	4115	4131	4147	4162	4177
1.4	0.4192	4207	4222	4236	4251	4265	4279	4292	4306	4319
1.5	0.4332	4345	4357	4370	4382	4394	4406	4418	4429	4441
1.6	0.4452	4463	4474	4484	4495	4505	4515	4525	4535	4545
1.7	0.4554	4564	4573	4582	4591	4599	4608	4616	4625	4633
1.8	0.4641	4649	4656	4664	4671	4678	4686	4693	4699	4706
1.9	0.4713	4719	4726	4732	4738	4744	4750	4756	4761	4767
2.0	0.4772	4778	4783	4788	4793	4798	4803	4808	4812	4817
2.1	0.4821	4826	4830	4834	4838	4842	4846	4850	4854	4857
2.2	0.4861	4864	4868	4871	4875	4878	4881	4884	4887	4890
2.3	0.4893	4896	4898	4901	4904	4906	4909	4911	4913	4916
2.4	0.4918	4920	4922	4925	4927	4929	4931	4932	4934	4936
2.5	0.4938	4940	4941	4943	4945	4946	4948	4949	4951	4952
2.6	0.4953	4955	4956	4957	4959	4960	4961	4962	4963	4964
2.7	0.4965	4966	4967	4968	4969	4970	4971	4972	4973	4974
2.8	0.4974	4975	4976	4977	4977	4978	4979	4979	4980	4981
2.9	0.4981	4982	4982	4983	4984	4984	4985	4985	4986	4986
3.0	0.4987	4990	4993	4995	4997	4998	4998	4999	4999	5000

TABLE 4. *The t-distribution**

d.f.	P = 0.1	0.05	0.02	0.01	0.001
1	6.314	12.706	31.821	63.657	636.619
2	2.920	4.303	6.965	9.925	31.598
3	2.353	3.182	4.541	5.841	12.924
4	2.132	2.776	3.747	4.604	8.610
5	2.015	2.571	3.365	4.032	6.859
6	1.943	2.447	3.143	3.707	5.959
7	1.895	2.365	2.998	3.499	5.408
8	1.860	2.306	2.896	3.355	5.041
9	1.833	2.262	2.821	3.250	4.781
10	1.812	2.228	2.764	3.169	4.587
11	1.796	2.201	2.718	3.106	4.437
12	1.782	2.179	2.681	3.055	4.318
13	1.771	2.160	2.650	3.012	4.221
14	1.761	2.145	2.624	2.977	4.140
15	1.753	2.131	2.602	2.947	4.073
16	1.746	2.120	2.583	2.921	4.015
17	1.740	2.110	2.567	2.898	3.965
18	1.734	2.101	2.552	2.878	3.922
19	1.729	2.093	2.539	2.861	3.883
20	1.725	2.086	2.528	2.845	3.850
21	1.721	2.080	2.518	2.831	3.819
22	1.717	2.074	2.508	2.819	3.792
23	1.714	2.069	2.500	2.807	3.767
24	1.711	2.064	2.492	2.797	3.745
25	1.708	2.060	2.485	2.787	3.725
26	1.706	2.056	2.479	2.779	3.707
27	1.703	2.052	2.473	2.771	3.690
28	1.701	2.048	2.467	2.763	3.674
29	1.699	2.045	2.462	2.756	3.659
30	1.697	2.042	2.457	2.750	3.646
40	1.684	2.021	2.423	2.704	3.551
60	1.671	2.000	2.390	2.660	3.460
120	1.658	1.980	2.358	2.617	3.373
∞	1.645	1.960	2.326	2.576	3.291

* Abridged from Table 3 of Fisher and Yates,
*Statistical Tables for Biological, Agricultural and
Medical Research*, Oliver & Boyd, Edinburgh,
by permission of the publishers.

TABLE 5. *Random numbers*

19	90	70	99	00	20	21	14	68	86	14	52	41	52	48	87	63	93	95	17	11	29	01	95	80
65	97	38	20	46	85	43	01	72	73	03	37	18	39	11	08	61	74	51	69	89	74	39	82	15
51	67	11	52	49	59	97	50	99	52	18	16	36	78	86	08	52	85	08	40	87	80	61	65	31
17	95	70	45	80	72	68	49	29	31	56	80	30	19	44	89	85	84	46	06	59	73	19	85	23
63	52	52	01	41	88	02	84	27	83	78	35	34	08	72	42	29	72	23	19	66	56	45	65	79
60	61	97	22	61	49	64	92	85	44	01	64	18	39	96	16	40	12	89	88	50	14	49	81	06
98	99	46	50	47	12	83	11	41	16	63	14	52	32	52	25	58	19	68	70	77	02	54	00	52
76	38	03	29	63	79	44	61	40	15	86	63	59	80	02	14	53	40	65	39	27	31	58	50	28
53	05	70	53	30	38	30	06	38	21	01	47	59	38	00	14	47	47	07	26	54	96	87	53	32
02	87	40	41	45	47	24	49	57	74	22	13	88	83	34	32	25	43	62	17	10	97	11	69	84
35	14	97	35	33	68	95	23	92	35	56	54	29	56	93	87	02	22	57	51	61	09	43	95	06
94	51	33	41	67	13	79	93	37	55	14	44	99	81	07	39	77	32	77	09	85	52	05	30	62
91	52	80	32	44	09	61	87	25	21	13	80	55	62	54	28	06	24	25	93	16	71	13	59	78
65	09	29	75	63	20	44	90	32	64	53	89	74	60	41	97	67	63	99	61	46	38	03	93	22
20	71	53	20	25	73	37	32	04	05	56	07	93	89	30	69	30	16	09	05	88	69	58	28	99
01	82	77	45	12	07	10	63	76	35	19	48	56	27	44	87	03	04	79	88	08	13	13	85	51
53	43	37	15	26	92	38	70	96	92	82	11	08	95	97	52	06	79	79	45	82	63	18	27	44
11	39	03	34	25	99	53	93	61	28	88	12	57	21	77	52	70	05	48	34	56	65	05	61	86
40	36	40	96	76	93	86	52	77	65	99	82	93	24	98	15	33	59	05	28	22	87	26	07	47
99	63	22	32	98	18	46	23	34	27	43	11	71	99	31	85	13	99	24	44	49	18	09	79	49
58	24	82	03	47	24	53	63	94	09	74	54	13	26	94	41	10	76	47	91	44	04	95	49	66
47	83	51	62	74	22	06	34	72	52	04	32	92	08	09	82	21	15	65	20	33	29	94	71	11
23	05	47	47	25	07	16	39	33	66	18	55	63	77	09	98	56	10	56	79	77	21	30	27	12
69	81	21	99	21	29	70	83	63	51	70	47	14	54	36	99	74	20	52	36	87	09	41	15	09
35	07	44	75	47	57	90	12	02	07	54	96	09	11	06	23	47	37	17	31	54	08	01	88	63
55	34	57	72	69	33	35	72	67	47	82	80	84	25	39	77	34	55	45	70	08	18	27	38	90
69	66	92	19	09	49	41	31	06	70	05	98	90	07	35	42	38	06	45	18	64	84	73	31	65
90	92	10	70	80	65	19	69	02	83	67	72	16	42	79	60	75	86	90	68	24	64	19	35	51
86	26	98	29	06	92	29	84	38	76	63	49	30	21	30	22	60	27	69	85	29	81	94	78	70
74	16	32	23	02	98	77	87	68	07	66	39	67	98	60	91	51	67	62	44	40	98	05	93	78
39	40	04	59	81	00	41	86	79	79	47	53	53	38	09	68	47	22	00	20	35	55	31	51	51
15	91	29	12	03	57	99	99	90	37	75	91	12	81	19	36	63	32	08	58	37	40	13	68	97
90	49	22	23	62	12	59	52	57	02	55	65	79	78	07	22	07	90	47	03	28	14	11	30	79
98	60	16	03	03	31	51	10	96	46	54	34	81	85	35	92	06	88	07	77	56	11	50	81	69
39	41	88	92	10	96	11	83	44	80	03	92	18	66	75	34	68	35	48	77	33	42	40	90	60
16	95	86	70	75	85	47	04	66	08	00	83	26	91	03	34	72	57	59	13	82	43	80	46	15
52	53	37	97	15	72	82	32	99	90	06	66	24	12	27	63	95	73	76	63	89	73	44	99	05
56	61	87	39	12	91	36	74	43	53	13	29	54	19	28	30	82	13	54	00	78	45	63	98	35
21	94	47	90	12	77	53	84	46	47	85	72	13	49	21	31	91	18	95	58	24	16	74	11	53
23	32	65	41	18	37	27	47	39	19	65	65	80	39	07	84	83	70	07	48	53	21	40	06	71

TABLE 5 (*continued*)

00	83	63	22	55	34	18	04	52	35	99	01	30	98	64	56	27	09	24	86	61	85	53	83	45
87	64	81	07	83	11	20	99	45	18	45	76	08	64	27	48	13	93	55	34	18	37	79	49	90
20	69	22	40	98	27	37	83	28	71	69	62	03	42	73	00	06	41	41	74	45	89	09	39	84
40	23	72	51	39	10	65	81	92	59	73	42	37	11	61	58	76	17	14	97	04	76	62	16	17
73	96	53	97	86	59	71	74	17	32	64	63	91	08	25	27	55	10	24	19	23	71	82	13	74
38	26	61	70	04	33	73	99	19	87	95	60	78	46	75	26	72	39	27	67	53	77	57	68	93
48	67	26	43	18	87	14	77	43	96	99	17	43	48	76	43	00	65	98	50	45	60	33	01	07
55	03	36	67	68	72	87	08	62	40	24	62	01	61	16	16	06	10	89	20	23	21	34	74	97
44	10	13	85	57	73	96	07	94	52	19	59	50	88	92	09	65	90	77	47	25	76	16	19	33
95	06	79	88	54	79	96	23	53	10	48	03	45	15	22	65	39	07	16	29	45	33	02	43	70
68	15	54	35	02	42	35	48	96	32	95	33	95	22	00	18	74	72	00	18	22	85	61	68	90
58	42	36	72	24	58	37	52	18	51	90	84	60	79	80	24	36	59	87	38	67	80	43	79	33
95	67	47	29	83	94	69	40	06	07	46	40	62	98	82	54	97	20	56	95	27	62	50	96	72
98	57	07	23	69	65	95	39	69	58	20	31	89	03	43	38	46	82	68	72	33	78	80	87	15
56	69	47	07	41	90	22	91	07	12	71	59	73	05	50	08	22	23	71	77	13	13	92	66	99

Index

175

176   *Index*